T0186373

Systems Evaluation

Methods, Models, and Applications

Systems Evaluation, Prediction, and Decision-Making Series

Series Editor
Yi Lin, PhD
Professor of Systems Science and Economics
School of Economics and Management
Nanjing University of Aeronautics and Astronautics

Systems Evaluation: Methods, Models, and Applications
Sifeng Liu, Naiming Xie, Chaoqing Yuan, and Zhigeng Fang
Nanjing University of Aeronautics & Astronautics
ISBN: 978-1-4200-8846-5

Measurement Data Modeling and Parameter Estimation
Zhengming Wang, Dongyun Yi, Xiaojun Duan, Jing Yao, Defeng Gu
National University of Defense Technology, Changsha, PR of China
ISBN: 978-1-4398-5378-8

Optimization of Regional Industrial Structures and Applications
Yaoguo Dang, Sifeng Liu, and Yuhong Wang, Nanjing University of Aeronautics
and Astronautics
ISBN: 978-1-4200-8747-5

Hybrid Rough Sets and Applications in Uncertain Decision-Making
Lirong Jian, Sifeng Liu, and Yi Lin, Nanjing University of Aeronautics and Astronautics
ISBN: 978-1-4200-8748-2

Irregularities and Prediction of Major Disasters
Yi Lin, Nanjing University of Aeronautics and Astronautics
ISBN: 978-1-4200-8745-1

Theory and Approaches of Unascertained Group Decision-Making
Jianjun Zhu, Nanjing University of Aeronautics and Astronautics
ISBN: 978-1-4200-8750-5

Theory of Science and Technology Transfer and Applications
Sifeng Liu, Zhigeng Fang, Hongxing Shi, and Benhai Guo,
Nanjing University of Aeronautics and Astronautics
ISBN: 978-1-4200-8741-3

Grey Game Theory and Its Applications in Economic Decision-Making
Zhigeng Fang, Nanjing University of Aeronautics and Astronautics
ISBN: 978-1-4200-8739-0

Systems Evaluation

Methods, Models, and Applications

Sifeng Liu · Naiming Xie

Chaoqing Yuan · Zhigeng Fang

CRC Press
Taylor & Francis Group
Boca Raton London New York

CRC Press is an imprint of the
Taylor & Francis Group, an **informa** business

CRC Press
Taylor & Francis Group
6000 Broken Sound Parkway NW, Suite 300
Boca Raton, FL 33487-2742

Printed in the United States of America on acid-free paper
Version Date: 20111102

International Standard Book Number: 978-1-4200-8846-5 (Hardback)

Library of Congress Cataloging-in-Publication Data

Systems evaluation : methods, models, and applications / Sifeng Liu ... [et al.].
    p. cm. -- (Systems evaluation, prediction, and decision-making series)
  Includes bibliographical references and index.
  ISBN 978-1-4200-8846-5 (hbk. : alk. paper)
  1. System analysis. 2. Decision making. I. Liu, Sifeng.

QA402.S975 2012
003--dc23   2011039594

Visit the Taylor & Francis Web site at
http://www.taylorandfrancis.com

and the CRC Press Web site at
http://www.crcpress.com

Contents

Preface...xiii

Acknowledgments xv

Introduction...xvii

Authors...xix

1 Common System Evaluation Methods and Models............................1
   1.1   Introduction 1
   1.2   Qualitative Evaluation..3
        1.2.1   A Summary on Qualitative Evaluation 3
        1.2.2   Nominal Group Technique 5
        1.2.3   Brainstorming 5
        1.2.4   Delphi 8
                1.2.4.1   Features 9
                1.2.4.2   Methods Derived from Delphi 9
                1.2.4.3   Preparing Questionnaires 12
                1.2.4.4   Evaluation Process 15
                1.2.4.5   Delphi Principles to Observe 15
   1.3   Index System for Evaluation 17
        1.3.1   Determining Structure of Index System 19
        1.3.2   Statistical Analysis of Indicator Information.........20
        1.3.3   Determining Values of Indicators...........................21
                1.3.3.1   Quantified Indicators 21
                1.3.3.2   Normalization 22
        1.3.4   Determining Indicator Weights..............................23
                1.3.4.1   Least Squares Method 23
                1.3.4.2   Eigenvector Method 23
                1.3.4.3   Other Weighting Methods 24

1.4 Comparative Evaluation and Logical Framework Approach............24
    1.4.1 Comparison Method for Main Indicators...........................25
        1.4.1.1 Method 1..25
        1.4.1.2 Method 2..25
    1.4.2 Logical Framework Approach..26
        1.4.2.1 Basic Concept..26
        1.4.2.2 Goal Levels...27
        1.4.2.3 Vertical Logic of Causality 28
        1.4.2.4 Horizontal Logic 29
        1.4.2.5 Verifying Indicators..29
        1.4.2.6 Verification Method 30
        1.4.2.7 Logical Framework Approach Used in
                Postevaluation...30
1.5 Analytic Hierarchy Process...32
    1.5.1 Hierarchical Structuring...32
    1.5.2 Constructing Judgment Matrix 33
    1.5.3 Single Level Ordering and Consistency Check 34
    1.5.4 Hierarchy Total Ordering and Consistency Check.........36
1.6 DEA Relative Efficiency Evaluation...39
    1.6.1 DEA Efficiency Evaluation Index and C^2R Model...............39
        1.6.1.1 DEA Efficiency Evaluation Index 39
        1.6.1.2 C^2R Design 40
    1.6.2 DEA Validity Judgment 42
    1.6.3 Improving DEA Validity 43
1.7 Chapter Summary 44

2 Grey System Evaluation Models 45
2.1 Introduction 45
2.2 Generalized Grey Incidences Model 46
    2.2.1 Absolute Grey Incidence Model..................................... 46
    2.2.2 Relative Grey Incidence Model.......................................50
    2.2.3 Synthetic Grey Incidence Model....................................52
2.3 Grey Incidence Models Based on Similarity and Nearness 53
2.4 Grey Evaluation Using Triangular Whitenization Functions..........54
    2.4.1 Triangular Whitenization Function
        of Moderate Measure...55
    2.4.2 Evaluation Model Using End-Point Triangular
        Whitenization Functions 56
    2.4.3 Evaluation Model Using Center-Point Triangular
        Whitenization Functions..57
2.5 Multiattribute Grey Target Decision Model 58
    2.5.1 Basic Concepts 58
    2.5.2 Construction of Matrix of Uniform Effect Measures.........60

2.5.3 Construction of Matrix of Synthetic Effect Measures.........63
2.5.4 Steps of Multiattribute Grey Target
Assessment Algorithm 64

**3 Postevaluation of Road–Bridge Construction: Case Study of
Lianxu Highway in China..67**
3.1 Introduction 67
3.1.1 Postevaluation 67
3.1.1.1 Comparison of Feasibility Evaluation
and Preevaluation 68
3.1.1.2 Comparison with Midevaluation 68
3.1.1.3 Comparison of Acceptance and Audit 69
3.1.2 Lianxu Highway Project...69
3.1.2.1 Overview of Project 69
3.1.2.2 Design Parameters 71
3.2 Process Evaluation 71
3.2.1 Preliminary Work and Evaluation 71
3.2.2 Process Design 72
3.2.2.1 Blueprint Design and General Information 72
3.2.2.2 Preparation of Tender Documents.....................73
3.2.2.3 Project Implementation and Start of
Construction 73
3.2.2.4 Main Technical Indicators and
Evaluation of Changes...............................73
3.2.3 Implementation and Evaluation of Investment 73
3.2.3.1 Investment Changes 73
3.2.3.2 Analysis of Investment Changes 76
3.2.3.3 Financing Options.............................. 77
3.2.3.4 Analysis of Financing Costs............................. 77
3.2.4 Operating Conditions and Evaluation 77
3.2.4.1 Forecast and Evaluation of Traffic 77
3.2.4.2 Analysis of Vehicle Speed.............................79
3.2.4.3 Evaluation of Structural Changes in Traffic........79
3.2.4.4 Evaluation of Traffic Safety Management...........79
3.2.5 Evaluation of Management, Support, and
Service Facilities 80
3.2.5.1 Management..80
3.2.5.2 Support and Service Facilities 81
3.3 Traffic Forecasting...81
3.3.1 Basis 81
3.3.2 Forecasting 82
3.3.2.1 Based on Trend of High-Speed Flow
to Forecast 82

	3.3.2.2	Based on Regional Transportation System High-Speed Flow to Forecast	82
	3.3.2.3	Forecasting Impacts of Ports on Highway Traffic	82
	3.3.2.4	Forecasting Impacts of New Roads	82

3.4 Financial and Economic Evaluation 84
    3.4.1 Financial Evaluation 84
        3.4.1.1 Main Parameters 84
        3.4.1.2 Revenue and Costs 85
        3.4.1.3 Financial Evaluation 85
    3.4.2 Economic Evaluation 89
    3.4.3 Comparisons of Feasibility Study and Postevaluation 90
        3.4.3.1 Comparison of Financial Benefits 90
        3.4.3.2 Comparative Analysis of Economic Benefits 92
3.5 Assessment of Environmental and Social Impacts 93
    3.5.1 Environmental Impacts 93
        3.5.1.1 Evaluation of Environmental Management 93
        3.5.1.2 Implementation of Environmental Protection Measures 94
        3.5.1.3 Conclusions of Survey 94
    3.5.2 Social Impacts 94
        3.5.2.1 Division of Area of Coverage 94
        3.5.2.2 Economic Development Correlation of Highway and Line Side Areas 94
    3.5.3 Economic Development 95
    3.5.4 Macroeconomic Impact Analysis 96
        3.5.4.1 Impact on Total Economy 96
        3.5.4.2 Effects on Economic Structure 97
        3.5.4.3 Effects on Environment and Society 97
        3.5.4.4 Local Compatibility Analysis 97
3.6 Sustainability Evaluation of Project Objective 97
    3.6.1 Effects of External Conditions 97
        3.6.1.1 Socioeconomic Development 97
        3.6.1.2 Highway Network Development 97
        3.6.1.3 Transportation Development 97
        3.6.1.4 Management System 97
        3.6.1.5 Policies and Regulations 98
        3.6.1.6 Supporting Facilities 98
    3.6.2 Effects of Internal Conditions 98
        3.6.2.1 Operating Mechanism 98
        3.6.2.2 Internal Management 98
        3.6.2.3 Service Status 98

3.6.2.4 Impacts of Highway Tolls 98
3.6.2.5 Impacts of Operation Conditions 98
3.6.2.6 Impacts of Construction Quality.......................98
3.6.3 Comprehensive Evaluation of Sustainability.......................99
3.6.3.1 Evaluation Index...99
3.6.3.2 Determination of Index Weight 99
3.6.3.3 Conclusion 99
3.6.4 Means for Realizing Sustainability 99
3.7 Problems and Recommendations..100
3.7.1 Problems..100
3.7.1.1 Defective Analysis of Effects on Transfer, Induced Traffic Volume, and Gap between Forecast and Actual Value 100
3.7.1.2 Inadequate Design Based on Engineering Survey of Local Roads: Increased Design Costs and Impacts on Progress 100
3.7.1.3 Inadequate Funding Programs and Municipal Matching Requirements 101
3.7.1.4 Unreasonable Standards for Service Facilities: Wastes of Resources 102
3.7.1.5 Preliminary Test Section....................................102
3.7.1.6 Inadequate Environmental Protection and Pollution 102
3.7.1.7 Land Waste Caused by Defective Selection Programs for Local Roads................................103
3.7.2 Recommendations 103

4 Efficiency Evaluations of Scientific and Technological Activities 105
4.1 Introduction 105
4.2 Allocation Structure and Use Efficiency Analysis of Research Expenditures...106
4.2.1 Allocation of Funds for Science and Technology in China...106
4.2.2 Conclusions and Recommendations 113
4.2.2.1 Conclusions and Problems.................................113
4.2.2.2 Considerations and Recommendations.............114
4.3 Efficiency Evaluation of University Scientific and Technological Activities Based on DEA Model 115
4.3.1 Index Selection 115
4.3.2 Evaluation 116
4.3.3 Evaluation of Use of University Research Funds by Region..118

4.4 Evaluation of Regional Scientific and Technological Strength:
Jiangsu Province 118
    4.4.1 Evaluation of Scientific and Technological Strength of
        China Provinces 121
        4.4.1.1 Devising Evaluation Index System...................121
        4.4.1.2 Explanations of Indices.............................121
        4.4.1.3 Concrete Values of Evaluation Indices
            for Scientific and Technological Strength
            of Provinces 123
        4.4.1.4 Grey Clustering Evaluation of
            Scientific and Technological Strength of
            China Provinces 124
    4.4.2 Evaluation of Scientific and Technological Strengths
        of Prefecture-Level Cities in Jiangsu Province...................133
        4.4.2.1 Designing Evaluation Index System.................133
        4.4.2.2 Explanations of Indices.............................133
        4.4.2.3 Concrete Values of Evaluation Indices.............135
        4.4.2.4 Grey Clustering Evaluation 136

5 Evaluation of Energy Saving in China 141
    5.1 Introduction 141
    5.2 Energy-Saving Effects of Technological Progress 142
        5.2.1 Extended Cobb–Douglas Production Function............142
        5.2.2 Data...144
        5.2.3 Empirical Research......................................148
        5.2.4 Conclusion 151
    5.3 Energy-Saving Effect of Industrial Restructuring...........151
        5.3.1 Decomposition of Energy Intensity 152
        5.3.2 Changes of Energy Intensity Caused by
            Industrial Structure 153
        5.3.3 Grey Linear Programming Model for Analyzing
            Industrial Restructuring Impact on Energy Saving..........154
        5.3.4 Industrial Restructuring................................158
        5.3.5 Conclusion 172
    5.4 Energy-Saving Effect of Development and Use of
        Nonfossil Energy 173
        5.4.1 Energy Consumption Structure...........................174
        5.4.2 Energy Consumption Structure Forecasting...............175
        5.4.3 Conclusion 176
    5.5 Evaluation of Energy Policy......................................176
        5.5.1 Model..179
            5.5.1.1 With and without Antitheses.......................179
            5.5.1.2 Linear Regression 180

      5.5.2   Data...180

      5.5.3   Energy-Saving Effects of Energy Policies 182

            5.5.3.1   With and without Antitheses........................182

            5.5.3.2   Linear Regression 184

  5.6   Conclusion 186

6 International Cooperation Project Selection....................................189

  6.1   Overview...189

  6.2   Demand for and Supply of International Cooperative
Key Technology in Jiangsu Province..191

      6.2.1   Analysis of Demand 191

      6.2.2   Analysis of Supply 193

      6.2.3   Selection of Key Technologies for International
Cooperation by Leading Regional Industries 195

  6.3   Demand and Supply Indices of International
Cooperative Key Technology 196

      6.3.1   The Principle of Index System Construction....................196

            6.3.1.1   Integrity...197

            6.3.1.2   Scientific Rationality 197

            6.3.1.3   Index System 197

            6.3.1.4   Independence 197

            6.3.1.5   Feasibility 197

            6.3.1.6   Comparability 197

      6.3.2   Evaluating Indices of Urgency and Possibility 197

      6.3.3   Weights of Urgency and Possibility Indices 201

            6.3.3.1   Determining Weights 201

            6.3.3.2   Analysis of Urgency and Possibility
Index Weights...201

            6.3.3.3   Relative Importance of Urgency and
Possibility Indices 202

  6.4   Choosing Model Based on Grey Clustering with Fixed Weight.....203

      6.4.1   Methodology..204

      6.4.2   Priority Orders of Key Technology 208

  6.5   Selection of International Cooperative Key Technology 210

References 231

Index 235

Preface

Evaluation is a practical activity by which people judge the value of an object or process. The need for socioeconomic development and scientific decision making led to a wide range of evaluation activities such as engineering project evaluation, science project evaluation, industrial development assessment, environmental assessment, evaluation of university subject construction, enterprise competitiveness assessment, personnel quality evaluation related to technology and economic development evaluation, and even evaluation of comprehensive national strength and government policies. Evaluation is related to all aspects of production and life.

This book outlines various systems evaluation methods and models. The evolution of systems evaluation is presented in a clear, logical way, starting with qualitative assessment and proceeding to a description of the process and methods to building an index system of evaluation, and again with some common evaluation methods and models of comparative evaluation methods, the logical framework approach, analytic hierarchy process (AHP), and data envelopment analysis (DEA) methods. Several new evaluation models of grey systems including the general grey incidence model, grey incidence models based on similarity and closeness, grey cluster evaluation based on triangular whitenization functions, and multi-attribute grey target decision models are introduced. Empirical studies based on reality are introduced and cover evaluation of road–bridge construction projects, the efficiency evaluation of science and technology activities, assessment of energy saving in China, and the evaluation and selection of international cooperation projects.

This book is unique in its emphasis on the practical applications of systems evaluation methods and models. The methods and models are introduced briefly and we attempt to explain intricate concepts in an easily understandable way. In addition, practical examples illustrate the practical application, analysis, and computation of systems evaluation methods and models.

Chapters 1 and 2 are written by Sifeng Liu, Chapter 3 by Zhigeng Fang, Chapters 4 and 6 by Naiming Xie, and Chapter 5 by Chaoqing Yuan. Lirong Jian, Hongzhuan Chen, Jeffrey Forrest, Yaoguo Dang, Hecheng Wu, Chuanmin Mi, Shawei He, Yong Liu, Yaping Li, Zhaowen Shi, Jianfei Xiao, Lifang He, Ying Cao,

Liang Yu, Yuqiang Guo, Xiao Tang, Mei Wang, Tong Yin, Hongchang Ren, Pengtao Lin, Dufang Fan, Fei Wang and Xin Jin participated in related studies. Professor Sifeng Liu took charge of the draft summarization and the final approval.

This book is intended to serve as a textbook for postgraduates and senior undergraduate students specializing in economics and management and as a reference text for those engaged in management, scientific research, engineering technology, and other disciplines.

Any errors or omissions that may be pointed out by readers will be appreciated.

Acknowledgments

The relevant research for this book was supported by the National Natural Science Foundation of China (90924022, 70971064, 70901041, and 70701017), the Social Science Foundation of China (10zd&014 and 08AJY024), the Soft Science Foundation of China (2008GXS5D115), the Foundation for Doctoral Programs (200802870020 and 20093218120032), and the Foundation for Humanities and Social Sciences of the Ministry of Education (08JA630039). The authors also acknowledge the support of the Science Foundation of Jiangsu Province (Y0553-091), the Foundation for Key Research in Philosophy and Social Science of the Colleges and Universities of Jiangsu Province, and the Foundation for National Outstanding Teaching Group of China (10td128).

The authors consulted several experts, studied the research of many scholars, and received a great deal of help from Professor Jeffrey Forrest. We wish to thank them all.

Introduction

This book outlines systems evaluation methods and models. The qualitative assessment methods of nominal group technique, brainstorming, and Delphi are presented. The methods for building an evaluation index system, common evaluation methods and models of comparative evaluation, the logical framework approach, the analytic hierarchy process (AHP), and data envelopment analysis (DEA) for evaluating relative efficiency are presented as well. Several new grey systems evaluation models including the generalized grey incidence model, models based on different visual angles of similarity and/or closeness, cluster evaluation models based on end point and center point triangular whitenization functions, and multi-attribute target decision models are introduced. We cover empirical studies based on reality including postevaluation of road–bridge construction projects, efficiency evaluations of scientific and technological activities, the evaluation of energy saving efforts in China, and the evaluation and selection of international cooperation projects for Jiangsu Province.

This book is unique in its emphasis on the practical application of systems evaluation methods and models. We attempt to explain intricate concepts in an easily understandable way utilize step-by-step explanations of methods and models. The book is suitable as a text for postgraduates and senior undergraduate students specializing in economics and management. It is also intended as a reference for those interested in the methods and technology of complex assessments.

Authors

Sifeng Liu earned a B.S. in mathematics in 1981 from Henan University, an M.S. in economics in 1986, and a Ph.D. in systems engineering in 1998 from Huazhong University of Science and Technology, both in China. He served as a visiting professor at Slippery Rock University in Pennsylvania, De Montfort University in England, and Sydney University in Australia. At present, he is the director of the Institute for Grey Systems Studies and the dean of the College of Economics and Management at Nanjing University of Aeronautics and Astronautics (NUAA) and a distinguished professor, academic leader, and doctor tutor in management science and systems engineering. His research focuses on grey systems theory, systems evaluation, and prediction modeling. He has published over 200 research papers and 19 books.

Over the years, Dr. Liu received 18 provincial and national prizes for his outstanding achievements in scientific research and he was named a distinguished professor and an expert who made significant contributions to China. In 2002, he was recognized by the World Organization of Systems and Cybernetics. He is a member of the Evaluation Committee and the Teaching Direct Committee for Management Science and Engineering of the National Science Foundation of China (NSFC). Additionally, he currently serves as chair of the TC of IEEE SMC on Grey Systems, president of the Grey Systems Society of China (GSSC), vice president of the Chinese Society for Optimization, Overall Planning and Economic Mathematics (CSOOPEM), cochair of the Beijing and Nanjing chapters of IEEE SMC, vice president of the Econometrics and Management Science Society of Jiangsu Province (EMSSJS), and vice president of the Systems Engineering Society of Jiangsu Province (SESJS). He is the editor of *Grey Systems: Theory and Application* (Emerald, UK) and serves on the editorial boards of several professional journals including the *Journal of Grey Systems, Scientific Inquiry, Chinese Journal of Management Science, Systems Theory and Applications, Systems Science and Comprehensive Studies in Agriculture*, and the *Journal of Nanjing University of Aeronautics and Astronautics*. Dr. Liu was named a National Excellent Teacher in 1995, an Expert Enjoying Government's Special Allowance in 2000, a National Expert in 1998, and an Outstanding Managerial Person of China in 2005.

Naiming Xie earned a B.S. in 2003, M.S. in 2006, and Ph.D. in 2008 in grey systems theory at Nanjing University of Aeronautics and Astronautics in China. His research interests include grey systems theory and management science. He is an associate professor of the College of Economics and Management at the university. He also serves as the secretary-general of the Grey System Society of China (GSSC) and published 2 books and more than 20 papers.

Chaoqing Yuan earned a B.S. in 2000, M.S. in 2004, and Ph.D. in 2010 in economics and management science from the Nanjing University of Aeronautics and Astronautics in China. He is an associate professor in the College of Economics and Management at the university and is a member of the Grey System Society of China (GSSC). His main research interests are technological innovation management and energy policy. He has published 11 papers.

Zhigeng Fang earned a master's in management science from Xi'an Science and Engineering University in 1996 and a Ph.D. in management science and engineering from the Nanjing University of Aeronautics and Astronautics in 2006. He is a professor and the deputy director of the Institute for Grey Systems Studies and also assistant dean of the College of Economics at the Nanjing University of Aeronautics and Astronautics. He is a vice president of Grey Systems Society of China (GSSC), director of the Chinese Society of Optimization, Overall Planning and Economic Mathematics, executive director of the Complex Systems Research Committee, and deputy director of Jiangsu Postevaluation Research Center, and a member of the Services Science Global and IEEE Intelligent Transportation Systems Council. Fang's main research interests are project management, post-project evaluation, and grey systems. He completed 26 academic and research projects for national, provincial, and municipal departments and published over 80 research papers and books. He has received several provincial and national prizes for his outstanding achievements in scientific research and applications.

Chapter 1

Common System Evaluation Methods and Models

1.1 Introduction

Evaluation is a practical method for judging the value of an object or activity. It is a necessary activity of governments, commerce, and life and has been used throughout the evolution of the human race.

When choosing habitats, the first humans had to evaluate possible threats correctly to ensure survival. When our ancestors exchanged goods and bartered services, they needed to evaluate the offerings of both parties. Since the 1950s, socioeconomic development and scientific decision making led to a wide range of evaluation activities in all areas including engineering projects, science and research, industrial development, environmental assessment, evaluation of university subject matter, enterprise competitiveness, personnel quality evaluation, technology and economic development evaluation, and even evaluation of comprehensive national strength and government policies. All aspects of life and work involve evaluation. As our understanding of nature and human society deepens, we must evaluate complex realities to promote the rapid development of methods for evaluating a single target criterion, multiple criteria, qualitative and quantitative methods, static and dynamic systems, certainties and uncertainties, and evolution of individuals and groups.

The first evaluation method cited in the scholarly literature was Francis Edgeworth's work on statistical testing published in the *Journal of the Royal*

Statistical Society in 1888. This paper discussed weighting the different parts of an examination and introduced weighting into the evaluation process. Edgeworth is considered the creator of modern science evaluation. In 1913, Spearman published "Correlations of Sums or Differences" to explain the significance of weighting and promoted weighted thinking. In the mid-20th century, with the advent of multiple index evaluation methods, nondimensional indicators are now used to determine the weighted averages of quantitative values. Since the 1970s, research led to development of a variety of evaluation methods such as the multidimensional preference analysis of linear programming, the analytic hierarchy process (AHP), and data envelopment analysis (DEA). In the 1980s, the traditional study methods expanded to allow combinations of methods involving new ideas such as statistical theory, fuzzy theory, grey system theory, and information theory. The combination of fuzzy thinking and discriminating analysis and cluster analysis of multivariate statistics resulted in fuzzy pattern recognition and fuzzy clustering; the matter-element theory produced a fuzzy matter-element model; neural networks produced a fuzzy system based on neural networks; and a combination with AHP produced fuzzy AHP. Since the late 1990s, advances led to a variety of certainty and uncertainty evaluation methods and new features such as fuzzy grey matter element systems, grey rough model combined uses, and other refinements.

The rapid development of software further broadened and deepened the applications of various evaluation methods. Evaluation activities permeate every aspect of life. Large complex systems require new evaluation methods. System evaluation methods make comprehensive use of systems engineering principles, methods, models, and technologies to assess technical, economic, social, ecological, and other issues based on preset objectives. The steps of system evaluation are:

Step 1: Establish clear overall objectives.
Step 2: Analyze the existing system environment and constraints.
Step 3: Determine the evaluation mission.
Step 4: Establish evaluation index objectives.
Step 5: Select evaluation methods.
Step 6: Collect data.
Step 7: Prepare a comprehensive evaluation.

Evaluation of a system involves starting and ending points and the phases are preassessment, assessment, postassessment, and tracking:

- Preassessment evaluation of a project occurs during the feasibility study stage, before the system exists. Preassessment involves existing data or utilizes simulations and prediction data. It may require expert consultations and comprehensive evaluation of qualitative judgments.
- Assessment examines whether the predetermined goals and plans were implemented. Assessment may be ongoing during a project to detect problems

and determine whether the final product will meet design criteria. This step resolves implementation problems and indicates adjustments needed.

■ Postassessment is conducted to determine whether the intended target was achieved. It is an objective assessment of system performance and factors related to project completion. It may involve a survey of stakeholders to yield a qualitative assessment.

Tracking evaluation is performed after a project is in operation to assess its implementation, effectiveness, role, and impact. Tracking evaluation monitors project implementation and operation to determine whether the project targets and plan are reasonable and effective, whether efficiency targets are achieved, and identify reasons for success or failure. Other factors that may be investigated are lessons learned, feasibility of future projects, and effectiveness of management and decision making. Tracking evaluation can lead to recommendations for improvement and resolve implementation problems to improve investment efficiency.

This chapter introduces qualitative assessment and evaluation index objectives; it also offers a comparison of evaluation methods, the logical framework approach, AHP, DEA, and relative efficiency evaluation.

1.2 Qualitative Evaluation

Qualitative evaluation is used to evaluate the character, direction, and efficiency of staff based on their background and knowledge of the relevant activity. Qualitative analysis can be based on quantitative estimates characterized by:

■ Need for less data
■ Factors that cannot be quantified
■ Simplicity
■ Feasibility

Therefore, qualitative analysis is an indispensable factor of flexible evaluation. Qualitative evaluation can guide government and commercial operations to management people and projects and make decisions. Qualitative evaluation methods play an important role in economic development in China.

1.2.1 A Summary on Qualitative Evaluation

The first humans had no concepts of numbering systems and relied mainly on qualitative judgments to guide their behavior. They counted rocks in containers and measured rope lengths in attempts to quantify objects. Despite the great leaps arising from electronics and computer advances, qualitative judgments remain dominant. The ancient Greek philosopher Aristotle and the Chinese Taoist philosopher Lao Tzu used qualitative descriptions.

If data are insufficient, facts are not accurate, or situations are difficult to describe, quantitative analysis is useless and qualitative evaluation is an effective evaluation method. For example, large-scale public works projects often use qualitative evaluation to estimate social and economic effects because many factors are difficult to quantify. In addition, the financial turmoil of the stock markets and the impacts of interest rate changes have on the real estate market cannot be analyzed quantitatively analyzed and can be determined only via qualitative evaluation.

Qualitative study takes place on two levels. The first is not pure or lacks sufficient data; conclusions tend to be broad and speculative. The second is based on quantitative analysis at a higher level of qualitative evaluation. To enhance the credibility of findings, quantitative evaluation may be supplemented by qualitative evaluation. In the processes of development and modification, the qualitative and quantitative relationships of constraints can indicate clear boundaries and lead to qualitative decisions. Therefore, systematic reviews should involve qualitative analysis, and quantitative evaluation should be based on qualitative analysis. Qualitative analysis can make quantitative evaluation in-depth and specific. Although the quantitative evaluation can determine the main factors, it cannot quantify the remaining factors. Therefore, both quantitative and qualitative analyses are used to make the necessary adjustments to achieve rational evaluation findings.

Qualitative evaluation methods use induction to explore issues, understand activities and phenomena, human behavior, and views, and arrive at objective conclusions. Qualitative evaluation depends mainly on experience and judgment and is vulnerable to subjective factors. To improve the reliability of qualitative evaluation, we should pay attention to the following issues:

1. We should strengthen investigation to grasp the impacts of favorable and unfavorable conditions and other factors to achieve more realistic judgments.
2. Research should be conducted and data should be gathered to ensure that quantification is qualitative. Qualitative analysis is based on estimation and used to strengthen an evaluation.
3. Qualitative and quantitative evaluations should be combined to improve the quality of an evaluation and reveal the adjustments required for making a final decision. The combination should show the essential characteristics of the plan, improve the quality of the evaluation, and aid decision making.

The most common qualitative evaluation methods are expert consultations, group discussions, brainstorming, focus groups, the Delphi method, comparative evaluations, and logical frameworks. This section introduces focus groups, brainstorming, and the Delphi method. Comparative evaluation and the logical framework approach are described in Section 1.4. The general steps of qualitative evaluation are:

Step 1: Planning the evaluation.

Step 2: Determining the appropriate expertise and number of evaluators required. For different problems, select people familiar with the goal. This will improve the accuracy of the evaluation and achieve significant cost savings.

Step 3: Designing consultation interview. Use various comprehensive qualitative evaluation methods, design evaluation links and key points, and consider the views of evaluators.

Step 4: Collecting and analyzing opinions, compiling findings, and organizing vote.

Step 5: Analyzing vote to determine evaluation findings. If large differences of opinion arise, the third and fourth steps can be repeated until a convergence of views leads to consistent results.

1.2.2 Nominal Group Technique

The group discussion method is also known as the nominal group technique (NGT) and involves convening a group of experts (usually five to nine) to gradually form a consensus opinion of the evaluation method. The duration is usually 60 to 90 minutes and the following steps are followed:

Step 1: Plan the evaluation. The problem should be clear, precise, and concise.

Step 2: Select five to nine participants familiar with the problem.

Step 3: Design the survey outline. Comprehensive use of various qualitative evaluation methods, design evaluation links, and key points will allow the evaluators to judge the issues independently.

Step 4: Sort and classify the collected views and evaluate feedback.

Step 5: Hold a group forum to discuss individual opinions. All views should be treated equally (without emphasizing some or neglecting others) to avoid arguments among participants.

Step 6: Evaluate the importance of each item in a preliminary vote to assess the preferences of participants noted on feedback forms. List the items in the order of relative importance. Each participant should indicate his or her preference; group results are based on the average of individual judgments.

Step 7: Discuss the preliminary voting results. If some participants require more information, share the information with all members.

Step 8: Take a final vote.

1.2.3 Brainstorming

Brainstorming, advanced by A. F. Osborne in 1939, is a method for inspiring creative thinking. It plays an important role in most qualitative evaluation activities.

In the 1950s, brainstorming was widely used in evaluation, prediction, and decision making and remains popular. Brainstorming encourages the creative thinking by experts and involves the following principles:

- The issues to be decided should be limited. The terms used to propose assumptions should be strictly regulated so that participants focus only on the issues.
- Participants cannot raise doubts about or dismiss the views of others.
- Participants should be encouraged to improve and summarize assumptions that have been suggested to allow assumptions to be modified.
- Participants should ignore their personal and ideological concerns and thus create an atmosphere of freedom and positive motivation.
- The statement should be concise; detail is not required. Long statements impede development of a creative atmosphere.
- Participants are not allowed to discuss the recommendations.

Practical experience has shown that brainstorming and utilizing the creative thinking resulting from exchanges of information among experts can produce effective results within a short time. Several types of brainstorming include:

- Direct brainstorming follows certain rules, discusses a specific issue, and encourages collective creative activity. The rules include banning assessment of proposed ideas; limiting speaking times of participants; allowing participants to speak several times; compiling all ideas proposed.
- Doubtful brainstorming is a collective idea generation method. Two meetings are held at the same time. The first meeting complies with the principles of direct brainstorming. The second meeting involves doubting ideas proposed in the first meeting.
- Controlled brainstorming is mass production of ideas. Intellectual activity generates new ideas and is often used to develop long-term ideas.
- Encouraging observation is used to find a reasonable solution within certain limits.
- Strategy observation method can find unified solutions for specific issues.

To provide an environment of creative thinking, the number of participants and the time of meeting should be decided in advance. Appropriate group size is 10–15 people, and meeting time is 20–60 minutes. Participants are selected according to certain principles:

- If the participants know each other, they should be peers (same title and job level). Leaders should not take part because they will exert pressure on members.
- If the participants do not know each other, they may be selected from different job levels. Their occupations, backgrounds, and titles should not be

mentioned before or during the session. Members of the Academy of Sciences and masters should be treated equally; each member should receive contact information for other participants.

■ Reaching a consensus is not a necessary condition for group members. The group may wish to include experts who understand the issues.

The group organizer should:

■ Evaluate the issues and underlying problems and their causes and analyze the causes, and possible results (it is better to exaggerate the issue to make participants feel that conflict must be resolved).
■ Analyze domestic and international experiences with the problem and point out several solutions.
■ Formulate the problem based on the central issue and its subproblems. The problem should be simple and narrow.

It is best to leave organization of the brainstorming to the evaluation experts because understand scientific debate and are familiar with the use of brainstorming procedures. If the professional side of the issues is narrow, experts in assessment and evaluation should be asked to be responsible for the evaluation. Brainstorming groups are usually composed of (1) method scholars, who are experts in evaluation, prediction, and decision making; (2) idea producers, who are experts in discussion; (3) analysts, who are high-level experts in discussion and should evaluate the present situation and development trends; and (4) performers, who act on the results.

All participants should be capable of associative thinking. During brainstorming, a creative atmosphere should help them concentrate on the issues under discussion. Opinions and ideas may overlap. All results of brainstorming should be viewed as creations of a group. Some of the most valuable views are based on other opinions and represent composites of several ideas.

Sometimes participants would like to brainstorm in writing. If that is acceptable, they should be advised of the goal, useful ideas for solving the problem; proposed responses, and the plan proposed to resolve all the issues.

The statements of organizers should inspire participants to want to solve the issues quickly. Usually at the onset, organizers must ask participants to make brief statements to create an atmosphere of free exchange of views quickly and provoke discussion. The organizers are actively only at the beginning of the meeting, primarily to set the rules. After the participants are encouraged to act, new ideas will emerge. Greater numbers of views and ideas will generate more diverse views and a better probability of a viable resolution. Views that emerge during the meeting can be summarized using the following procedures:

1. List all views.
2. Explain each view in general terms.

3. Summarize each view.
4. Analyze duplicate and complementary views and prepare a list.
5. Form final evaluations.

Question brainstorming is often used for evaluation purposes. This approach involves directly doubting the views proposed. Doubting the views is a special process to evaluate the reasonableness of the conclusions. The participants should doubt and comprehensively review each view proposed, focusing on analysis and evaluation of possibly biased opinions. New ideas may arise as a result of questioning. Question brainstorming should follow the principles of direct brainstorming; it does not allow confirmation of the proposed views but still encourages new ideas.

During question brainstorming, organizers should first clarify the content of discussion, introduce the list after the merger of views, and ensure participants focus on the issues to achieve a comprehensive evaluation. The process continues until no view can be doubted. The final step is questioning the comments made during the process to form a final evaluation.

Experience shows that brainstorming can help avoid compromise solutions and uncover feasible solutions through continuous and fair analysis. Brainstorming is now widely used for military and civilian forecasts and evaluations. For example, in long-term technology planning at the U.S. Department of Defense, 50 experts were invited to a 2-week brainstorming meeting. Their task was to doubt a working paper submitted in advance and formulate the paper into a coherent report. After discussion, only 25 to 30% of the original working paper was preserved; this shows the value of brainstorming. In addition, the British Post Office, Lockheed, Coca-Cola, and the International Business Machines Corporation (IBM) actively conduct brainstorming sessions to make predictions and evaluations.

A feasible solution proposed by brainstorming cannot achieve the target queue or indicate the best way, so it should be supported by experts via collective evaluation. Their results should be statistically processed to obtain a comprehensive view.

1.2.4 Delphi

Delphi, developed by Rand Corporation in the 1940s, is used for technology assessment and forecasting. Delphi and the NGT approach discussed earlier are similar. Both methods aim to solve complicated decision problems by consulting and gathering views of evaluators; they exhibit differences as well. Evaluators who tend to use the Delphi method are usually more focused than those who use NGT. Delphi does not require personal contact; members respond in writing and can thus work from several locations. NGT can usually be completed within 1 to 2 hours; Delphi activities may take several months or longer.

Delphi is an offshoot of NGT. It involves sending an anonymous letter through several rounds of consultation to seek the views of experts. Opinions on each round of the organizers are collected, compiled, and then redistributed to the experts for analysis to develop new arguments. Because the method is repeated many times, the experts' opinions become more consistent and thus increase the reliability of the conclusions.

Delphi is a *systems analysis* approach involving opinions and value judgments. It breaks through the limitations of traditional quantitative analysis and leads to more reasonable and effective decision making . Based on the future development of the various possibilities and estimating the probability of future occurrences, Delphi offers a variety of choices for decision making. Other methods are not as effective for obtaining clear answers based on probability.

The next section explains several aspects of the Delphi method such as derived methods, selection of experts, survey preparation, evaluation, and principles to be observed.

1.2.4.1 Features

1. Anonymity: To overcome the psychological vulnerabilities inherent in meetings of experts, Delphi is anonymous. Experts invited to participate in evaluations do not know each other and this eliminates the influence of psychological factors. Experts evaluate results from previous rounds without having to modify or make their views public so their reputations are not affected.
2. Communication between rounds: Delphi is different from polling methods that typically involve four rounds. To preserve anonymity and disseminate the results to all participants, the organizers should distribute statistical evaluation results to all experts during each round of assessment.
3. Statistical properties of results: An important feature of the Delphi method is quantitative treatment of feedback from rounds. Delphi uses statistical methods to process the results.

1.2.4.2 Methods Derived from Delphi

Some scholars, through in-depth study of the Delphi method, modified the process and created a number of derivative methods divided into two categories: (1) to maintain the basic characteristics of the classical Delphi method derived methods; and (2) changing one or several features. The derived Delphi methods maintain the basic characteristics of the classic method but amend some parts to overcome deficiencies:

1. Provide an event list. The first round of the classic Delphi method provides experts with theme tables specific to the situation. This allows organizers to rule out preconceived ideas and allow individuals to exhibit their expertise. Some experts do not understand where to start and chaotic events cannot be summarized. Also it is difficult to ensure that experts in the first round meet

the organizers' requirements. To overcome these shortcomings, the organizers can master the relevant information, seek expert advice beforehand, or prepare an event list to be included in the first round letter to the experts. Of course, experts may be added later and may suggest modifications.

2. Provide experts with background detail. Participants in evaluations are generally experts in their fields and cannot be expected to know background such as domestic and international political and economic climates. It is thus necessary to provide relevant information to give experts a common starting point.

3. Reduce the number of rounds. Short-term experiments showed coordination after only two rounds. More rounds can be used, based on experience and complexity of the evaluation.

4. Require experts to project realistic deadlines. The classic Delphi method often requires experts to note dates of implementation of evaluation activities. Impossible projections reduce success rates.

5. Expertise weights. Delphi experts may not specialize in the problem area and may work in relevant fields. Using a weighted average will improve Delphi method results.

6. Confidence probability index. Delphi events involve confidence factors that represent the statistical properties of a group response. Confidence factor is based on affirmative answers, that is, 100% minus the "never" or "never happens" responses equals confidence probability. For example, if 30% of the experts answered "never," the confidence probability of the event is 70%. Confidence probability is a useful statistic because it reveals positive and negative answers.

The changes to the Delphi method involve the characteristics of anonymity and feedback:

1. Partial lifting of anonymity. Anonymity can help develop strengths without outside support and opposition. Removing some or all anonymity during a Delphi activity can maintain the benefits and speed the process. Lifting anonymity allows some experts to clarify their arguments, express opinions if they wish to do so, and make inquiries if required.

2. Partial lifting of feedback. If feedback is totally abolished, the second round results may be limited. Experimental studies have shown that feedback may inhibit creativity, but eliminating all feedback will dilute the characteristics of the Delphi method. Partial lifting of the feedback helps experts move closer to the median to avoid new evaluations and arguments. The experts' opinions are confirmed in the final result.

3. Selection of experts. Delphi requires the establishment of a leading group responsible for developing the theme, preparing the event list, analyzing and processing the results and, more importantly, selecting the experts.

The Delphi method values the judgments of experts. If the invited experts who are to evaluate the theme do not have extensive knowledge of the subject, it will be difficult to obtain valid ideas and value judgments. Even if the evaluation topic is relatively narrow and highly targeted, it may be difficult to identify experts who have extensive knowledge of the topic areas. Therefore, identifying appropriate experts is one of the organizers' main tasks and the key to the success of Delphi.

The experts selected must agree to serve. It is not proper to send questionnaires to experts without their prior consent and some may decline participation. Statistics show that it is common to distribute 200 to 300 questionnaires in the first round and the response rate may be 50 percent or fewer. Thus, if questionnaires are distributed blindly without prior consent, it will be impossible to achieve sufficient responses for the evaluation.

To determine how experts should be selected, consider (1) the expertise required, (2) the method of choosing experts, and (3) the individual selections. When organizing an evaluation, the experts to be selected should have worked in the relevant field for at least a decade. The level of expertise is determined by the evaluation task. If the issue requires in-depth understanding of the history and technology developments related to the task, selection is relatively simple because sufficient data about experts is readily available. If the evaluation tasks are related to the development of specific technologies, it is preferable to select experts from both inside and outside the operation. To select outside experts:

1. List the issues that require expert response.
2. Based on the issues, determine the expertise needed and prepare a list of targeted questions.
3. Select experts, ask whether they can participate in the evaluation of the problem and send them lists of questions.
4. Determine the amount of time required by each participant and the funds needed to compensate them.

The selection of outside experts is difficult and generally takes several rounds of questions. The first list of experts will be familiar to the organizers. The experts in turn will identify other experts. The organizers will then send questionnaires those experts for advice, and ask them recommend one to two other experts. After that, the organizers select the experts from the list of participants based on recommendations and other factors.

Selection should include experts who are technically proficient and have stellar reputations along with those on the edge of their disciplines; sociology and economics experts may be included. Selecting experts who are responsible for technical leadership is important, but it is necessary to consider whether they have enough time to dedicate to answering questionnaires adequately. Experience has shown that experts in key positions rush to complete questionnaire and their opinions may not be as valuable as the organizers intended.

The number of experts depends on the problem size; generally 10 to 50 is a suitable number. Having too few experts results in limited representation of disciplines, lack of authority, and lack of reliability of evaluation results. Too many experts can make organizing difficult and produce more complicated results. However, for major problems, the number of experts can be extended to more than 100. Remember than even if experts agree to participate in an evaluation, they will not necessarily answer each round and drop out, so more experts than the number set at preselection should be chosen. After experts are selected, they should be divided into groups based on evaluation tasks and make appropriate group decisions.

Land Corporation in the United States used the Delphi method to evaluate and predict scientific breakthroughs, population growth, automation technology, and space technology and develop new weapons systems. The group consists of 82 members, divided into 6 specialized units. Half the members work for the company; the remainder includes a number of European.

A group of United States and Canada associations evaluated powder metallurgy technology and its impact on the forging industry. They organized an expert panel composed of 90 experts from raw material suppliers, equipment manufacturers, forging factories, forging users, and research institutes and divided the group into three units to evaluate growth trends of the forging of ferrous metal powders, nonferrous metals and high temperature metal powder forging, and cold forging.

The University of Manitoba invited 40 experts to evaluate and predict energy and environmental issues. Of the experts, 23 were from the United States, 8 from France, 3 from the United Kingdom, 2 from West Germany, 2 from Switzerland, and 1 each from Japan and Belgium. Each expert's median experience level was 21 years. Occupational distribution included 8 government workers, 6 academics, 11 employed by professional journals, and 15 working in industry and industrial research; 48% earned doctorates and 37% achieved master's level. About two-thirds of respondents participated in three rounds of evaluation, more than 30 experts participated in each round. The impacts of participant fluctuations were excluded.

1.2.4.3 Preparing Questionnaires

Before carrying out an assessment, a relevant questionnaire should be developed based on the evaluation tasks.

A. Use Questionnaire to Set Goals and Means (Table 1.1)

Organizers should analyze the available data to determine the overall objectives and subgoals. Experts may be invited to help establish the objectives and subgoals.

Means to achieve the objectives are based on research and development programs. Among many programs, choose a major one and ensure that other programs do not interfere. For example, when evaluating computer technology trends, an overall question is: "When human beings in all spheres of activity are accustomed to solving problems effectively by using computers, what is the computer technology development trend?" The subgoals of the question may be (1) solving human contact issues; (2) improving computer intelligence; (3) improving computer efficiency; and (4) enhancing the efficiency of installed capacity. Means for achieving the objectives including improvements of (a) cell technology; (b) peripheral equipment and communication technology; (c) information processing methods (mathematical models); (d) Cheng codes; (e) computer structures; (f) computer organization; and (g) computer design methods.

B. Formulate Questionnaires for Experts' Responses

Questionnaires are vital tools for Delphi because they provide information. The quality of questionnaires greatly affects the reliability of the results. Tabulated questionnaires such as Table 1.1 are very effective because they expedite experts' responses and allow organizers to classify responses. Table questions fall into three categories:

a. Quantitative estimate questionnaires can be used to determine completion times, technical parameters, probabilities, and interactions of various factors. Table 1.2 is a general questionnaire covering completion time of an event.
b. The answers to the example problem are broken down into three categories:
   i. Affirmative answer without conditions: For example, "How can we further improve computer production efficiency between 2012 and 2016?" Table 1.3 shows the general form of the questionnaire.
   ii. Deductive answer: For example, "To expand market share and improve profitability in 20121 and 2013, what measures should be taken?" Table 1.4 shows a general questionnaire.
   iii. Conditional answer: For example, "If some new theory is developed in the future, how do you think it will change computer structures?"

Table 1.1   Questionnaire Surveying Goals and Means

		Subgoal A	Subgoal B	Subgoal C	Subgoal D	Subgoal E	Subgoal F
Means of achieving objectives	Means a						
	Means b						
	Means c						
	Means d						
	Means e						
	Means f						
	Means g						
	Means h						

Table 1.2   Event Completion Time Questionnaire

	Completion Time		
	10% Probability	50% Probability	90% Probability
Solve science and technology problem	a_{1i}	b_{1i}	m_{1i}
Design machine	a_{2i}	b_{2i}	M_{2i}
Develop device with certain technical capabilities	a_{3i}	b_{3i}	m_{3i}

Note: Subscript numbers indicate incident completion sequences.

Table 1.3   Affirmative Answer Questionnaire

Measure	To Improve Product Quality and Increase Product Variety, which The Following Measures Would be the Most Effective?
Improve product structure	
Improve manufacturing process	
Add production capacity	

Table 1.4   Deductive Answer Questionnaire

Measure	To Expand Market Share and Improve Profitability, Which of the Following Measures Would be the Most Effective?
Increase advertising investment	
Improve product design and/or quality	
Reduce sale price	

   c. Full description questionnaires are of two kinds:
      i. They require list type answers. An example question is "What are the characteristics of a fifth-generation computer?"
      ii. They seek opinions. An example question is "How would you develop computer technology to establish a national computer network?"

1.2.4.4 Evaluation Process

After questionnaires are formulated, the evaluation can begin. Evaluation requires conditions that allow experts to judge ideas freely and independently. The classical Delphi evaluation has four rounds. In round 1, a survey listing the proposed evaluation tasks is sent to experts. Organizers collect and sort the returned surveys, merging similar incidents, excluding minor issues, and developing a precise list of terms. Then a second round of the questionnaire is sent to the experts. For example, the experts in round 1 in the example cited above involving growth trends in the forging industries cited more than 150 items; the leading group synthesized and presented 121.

In round 2, experts evaluate each event and explain their evaluations. Organizers analyze the data based on the experts' advice. Based on the results of round 2, the experts in round 3 make further judgments and again state their reasons. The third round may require only experts holding heretical opinions to adequately state their reasoning, because their reasons are often based on relevant external factors and may affect judgments of other members. In the fourth and final round, experts evaluate results from round 3; organizers may require some experts to restate their arguments.

After four rounds, the opinions of experts generally converge. For example, the Land Corporation project involved 50 pieces of evaluation and prediction for 49 events covering 6 issues. After 4 rounds, 31 events achieved fairly consistent results. The U.S. Society of Manufacturing Engineers in collaboration with the University of Michigan organized 125 and 150 experts, respectively, to evaluate various aspect of production and management technology. After four rounds, the results were very close; only 20% of the experts dissented.

1.2.4.5 Delphi Principles to Observe

Delphi will not meet all criteria in all cases. However, when applied to large analysis and research projects, the following principles apply:

1. Prepare a full description of the Delphi method. To make experts fully understand the situation, the questionnaire should include explanations of the purpose and tasks of evaluation and the roles of the experts along with a full description of the Delphi method. Because some experts may have misconceptions about Delphi, the organizers should clearly explain the substance and characteristics of the method, characteristics and the value of feedback from the experts.
2. Focus on issues. Issues should not be too scattered, so that the results constitute an organic whole. Questions should be graded, with the comprehensive issues covered first. Align problems similarly; start with simple ones and progress to complex issues. This progressive approach facilitates experts' answers.

3. Do not combine events. If an event includes two aspects and the experts agree on only one, answering becomes difficult. For example, some experts would have difficulty answering "Water extract of deuterium (heavy hydrogen) as the raw material of nuclear power stations to be completed by year _____." One may suggest a date of completion of a nuclear power plant but want to use tritium instead of deuterium as the raw material. If the expert noted a date, it would appear that he agreed with the use of deuterium; if he refused to answer, he would appear skeptical about building nuclear power plants. Thus, a questionnaire should not combine two possible events.

4. Semantics should be clear and explicit. Some experts often feel that surveys vaguely describe problems because the organizers improperly use technical terms and jargon, for example, "In which year will private households universally have remote access terminal devices?" *Universally* is a vague word that lacks quantitative concepts. Some experts believe that 50% is a universal indicator and cite a date based on that; others believe that 80% is the indicator and determine another date and their evaluation results may vary greatly. In practice, if evaluating the survey includes information about annual growth rates of installations of terminal equipment in private households, the expert opinions may be completely consistent. Terms such as *universal, broad,* and *normal* should be avoided.

5. Views of the leading group should not be imposed. During an evaluation, if opposing views are not sufficiently considered, the leading group may think its view is dominant and may try to impose it on the remainder of the team for including in the next round of evaluation. This dynamic will affect the results and their reliability will be questionable.

6. The questionnaire should be as simple as possible. It should help rather than hinder experts to make evaluations and focus on specific issues. A questionnaire should never be complex or confusion. Participants should be able to select choices from lists or fill in blank information. Space should be left for experts to clarify their views.

7. Limit the number of questions. The number of questions depends on the responses desired. If a problem requires only a simple answer, more experts may participate. If a problem is more complex and opposing views must be considered, responses will be fewer. Generally 25 questions represent a maximum. If a problem requires more than 50 questions, the leading group should seriously study whether the problem is too scattered and not on point.

8. Compensate experts appropriately. During the 19th century, few experts were compensated for participating in Delphi projects; that was found to affect their enthusiasm. an organization employing the Delphi method should adequately compensate participants for their help.

9. Consider the workload to be generated by the evaluation. If the number of group members is relatively small, the results may not justify the workload. For example, Table 1.5 shows processing times for a group composed of 10

Table 1.5 Workload for Each Survey Round

Survey Number	Quantity (Person/Hour)
1	22
2	20
3	20
4	10.5
5	3.5

experts conducting 5 rounds of evaluation. Each of the first three founds requires about 20 hours of processing time based on each expert's participation in each round, for an average time of 2 hours. In an evaluation involving 50 participants, the time for processing evaluation results for each round is about five times that shown in Table 1.5. If a project involves more than 100 staff members, a computer should be used to track participation.

10. Determine time intervals between rounds. The intervals vary based on the complexity of the evaluation. Most assessments require intervals of 3 to 6 weeks. Internet technology facilitates communications and intervals of 1 to 3 weeks are now common. Intervals between founds should be based on the complexity of the issue and the experts' interest.

The principles of the Delphi method are based on a long history of its use. They may not be followed strictly in some situations and this may affect the reliability of the results. However, by researching and abiding by these principles, organizers can make fewer mistakes and obtain effective evaluations.

1.3 Index System for Evaluation

Indicators measure system status. To evaluate a system, we must first establish the system used to control and measure uniformity, i.e., an evaluation index system. An evaluation index system should be scientifically and objectively as comprehensive as possible and consider various factors used to characterize the system and status of the main factors and related variables. Important qualitative factors also require appropriate indicators reflecting changes to avoid biased evaluation results. Certain principles should be followed:

1. Scientific principles require the selection of indicators, calculation methods, information collection steps, and other factors. All factors must have scientific bases. If the factors are subjective, the results will have little value.
2. Completeness. Evaluation should include complete characterization of all the main factors and their associated variables to avoid bias and omissions.

3. Hierarchy. A large system usually contains several subsystems that may be divided into sub-subsystems. Corresponding evaluation index systems should break down each level and ultimately form a multilevel index system.
4. Independence. The same levels of evaluation indicators should be relatively independent. If indicators of different situations at the same level are very similar, appropriate scientific methods are needed to evaluate them.
5. Comparability is the essence of evaluation. Only by using comparable indicators is it possible to make accurate comparisons. The comparability principle requires evaluation at different times and spaces to assess comparabilities of different objects. The scope and caliber of an evaluation must be consistent in relative numbers, proportions, and averages in order to evaluate comparability.
6. The operational principle. In establishing an evaluation index system, its feasibility should be fully considered. First, be brief and include the fewest possible number of indicators to reflect the general state of the system; second, make sure measurement is easy, the calculation method is easy to understand, and the required data is readily available. Third, indicators should be selected to be as representative as possible so that the evaluation is easy to understand. Include quantitative and qualitative indicators, static and dynamic indicators indices, absolute and relative indicators, and dialectic unity.

Figure 1.1 depicts construction of an indicator system that includes qualitative analysis of indicators including specific system objectives, indicators for

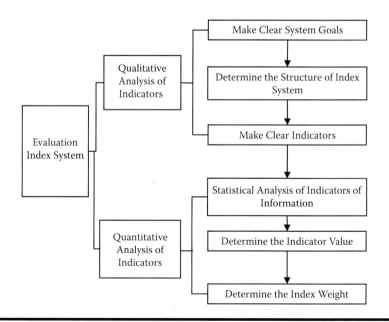

Figure 1.1   Construction of index system.

determining system architecture, and clear and specific indicator steps. After the initial target is set, quantitative analysis of data is used to determine target values and weights. The next section focuses on developing an evaluation index system.

1.3.1 Determining Structure of Index System

Many multiple attribute evaluation decisions are targeted at complex social and/ or economic systems in the fields of politics, economics, technology, ecology, and other disciplines. A wide range of subject matter and a lack of relevant information may create uncertainties in evaluation results. To solve complex, multilevel, and multifactor problems via a scientific evaluation, the first step is to analyze the system structure to identify interrelated and mutual restraint complex factors and determine their effects on the evaluation objectives. For factors involving only qualitative evaluation, appropriate and convenient quantification is needed. An index system should be based on a comprehensive evaluation of the system structure by developing preliminary indicators, collecting expert opinions, ensuring consistent exchanges of data via statistical processing and synthesis, and finally compiling a systematic evaluation index system that may have one of the following structures:

1. Single-layer system. All the individual indicators are at the same level, and each indicator can be evaluated by quantitative criteria; the structure is shown in Figure 1.2. Single-layer index systems are common in microeconomic management, for example, for selecting and buying equipment. Companies often evaluate technical and economic indicators with such systems.
2. Sequence-based system for multilevel index. All indicators can be decomposed into several lower level indices. The subindices may be broken down further until the lowest layer of indicators provides a quantitative evaluation based on specific criteria and indicators of different categories have no direct links. Figure 1.3 depicts this structure. This technique is used in macroeconomic management, for example, choosing a site for a manufacturing plant.
3. Non-sequence-based system for multilevel index. Some multilevel evaluations divide the subindices into several levels. The lowest level is the rule layer. Unlike the sequence-based method, non-sequence-based systems certain

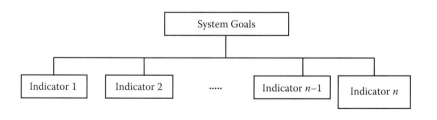

Figure 1.2   Single-layer evaluation system.

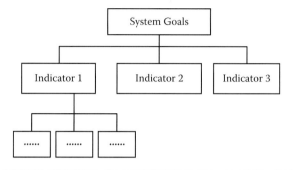

Figure 1.3   Sequence-based multilevel evaluation index system.

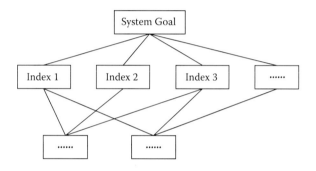

Figure 1.4   Nonsequential multilevel evaluation index system.

subindicators are not based on subindices at an adjacent level and cannot be cannot be classified by sequence relationships. See Figure 1.4. Nonsequential multilevel evaluation systems do not follow the independence principle.

1.3.2 Statistical Analysis of Indicator Information

An index system may be based on one of two methods: (1) subjective assessment and comparison of expert judgments; and (2) statistical analysis of data. The selection of an appropriate method should be based on information characteristics of the object and its environment.

1. The Delphi method based on the knowledge, wisdom, experience, intuition, reasoning, preferences, and values of members of an expert group is a typical type of expert assessment.
2. Statistical analysis based on principal component analysis selects and combines a few representative variables from a group of original variables by linear transformation. PCA is a dimensionality reduction process and is commonly used to select and simplify indicator systems. PCA indicators are independent of

each other and allow fewer indicators to reflect information represented by the larger group of original indicators. Let the original index vector (component) be $X = (X_1, X_2, \ldots X_n)$ and the new index vector (component) be $Y = (Y_1, Y_2, \ldots, Y_m)$; $Y_j, j = 1, 2, \ldots, m$ is a linear combination of $X_1, X_2, \ldots X_n$, that is

$$Y = CX^T \tag{1.1}$$

Y_1 has the greatest variance of all the linear combination of $X_1, X_2, \ldots, X_n$ and is called the first principal component. Y_2 has the second largest variance and is called the second principal component. C is the matrix composed of the eigenvector corresponding to the M eigenvalue ($\lambda_1 > \lambda_2 > \ldots > \lambda_m > 0$) of the characteristic equation

$$/R - \lambda I/ = 0 \tag{1.2}$$

where R is the correlation matrix after the standardization of the original target evaluation data.

$\lambda_j / \Sigma_{j=1}^m \lambda_j$, $j = 1, 2, \ldots, m$, is called the contribution rate of the j principal component. A higher contribution rate indicates that the corresponding principal component in the comprehensive evaluation is more important. $\Sigma_{j=1}^k \lambda_j / \Sigma_{j=1}^m \lambda_j$, $k = 1, 2, \cdots, m$ is called accumulative total contribution rate of the front k principal components. Cumulative contribution rate represents the size of the impact of the comprehensive evaluation of the first k components and may reveal the extent of the information contained in the original n indicators $X_1, X_2, \ldots, X_n$.

The final number of retained principal components must be based on the size of the cumulative contribution rate and the information revealed by the new index vector. Generally, if the first k principal components of the total contribution rate exceed 85%, the new index vector basically represents the original information contained by n indicators $X_1, X_2, \ldots, X_n$. If you request a new index vector containing more information, you can increase the standard of the total contribution rate, but the number of principal components selected may also increase. Conversely, you can reduce the standard of the cumulative contribution rate and the principal component number will decrease.

1.3.3 Determining Values of Indicators

This mainly refers to quantified and normalized data.

1.3.3.1 Quantified Indicators

This technique relies on scientific arguments and advice based on calculation. Specific steps to determine a quantified indicator are:

Step 1: Determine the mapping values to indicate the corresponding index.
Step 2: Select the calculation method and improvement factors.

Step 3: State the objective values of mapping variables.

Step 4: Determine the index evaluation method to evaluate the degree of realization of corresponding indicators based on concrete values of mapping variables.

To obtain quantified index data, consider the following parameters:

1. Position (level) is the overall average and a measure of central tendency of the evaluated object; the common formula involves the arithmetic and geometric means.

$$\bar{X}_j = \frac{1}{N} \sum_{i=1}^{N} X_{ij}, j = 1, 2, \cdots, M \tag{1.3}$$

$$\bar{X}_j = \sqrt[N]{\prod_{i=1}^{y} x_{ij}}, j = 1, 2, \cdots, M \tag{1.4}$$

where x_{ij} is the quantization value of j indicator of i evaluated object, N is the number of samples of evaluated object, and M is the number of evaluated indicators.

2. Discrete parameters show the dispersion of quantitative data of the evaluated index, such as the average deviation and standard deviation.

3. Distribution parameters. If the quantitative evaluation parameters show normal distribution, the standard deviation coefficient can reflect the symmetry. The formula is

$$G_j = \sqrt{\frac{1}{6N} \sum_{i=1}^{y} (\frac{\bar{X}_{ij} - X_{ij}}{\sigma_j})^3}, \, j = 1, 2, \cdots, M \tag{1.5}$$

If $G_j > 0$, there is a positive deviation; if $G_j = 0$, there is symmetry; and if $G_j < 0$, there is a negative deviation.

The digital features based on quantitative indicators can be modified or adjusted to a range of values. They can also serve as the basis for constructing the whitening of grey information and the membership function of fuzzy information.

1.3.3.2 Normalization

The main purpose of normalization is to solve the non-incommensurate index value with a unified value. The most common normalization method is the transformation and it is usually normalized to the value of the dimensionless 0–1 interval. Transformation is often divided into linear and nonlinear categories. Linear transformation is divided into standardized transformation, extreme transformation, mean transformation, initial transformation, and modular transformation methods. The following formula is the nonlinear transformation:

$$X'_{ij} = (\frac{X_{ij} - X_{\min j}}{X_{\max j} - X_{\min j}})^N \quad i = 1, 2, \cdots, N \quad j = 1, 2, \cdots, M \tag{1.6}$$

If $K = 1$, (1.6) becomes a linear limit transformation. In the formula, $X_{\min j}$ and $X_{\max j}$ are minimum and maximum of N evaluated sample values of the j indicator. If $K = 1$ $X_{\min j} = 0$, then (1.6) becomes a limit comparison formula.

1.3.4 Determining Indicator Weights

Weights are the important information of comprehensive evaluation. Weights should be determined based on the relative importance of indicators (namely, the contributions of indicators to comprehensive evaluation). Based on information infrastructure, you can choose the qualitative experience methods, precise quantitative data processing methods, and hybrid methods to determine the weights. The common feature of these methods is the *contrast;* whether the results of the contrast are accurate and consistent is the key to determining the reasonableness of the income weights. Commonly used contrast methods are weighted least squares and the eigenvector.

1.3.4.1 Least Squares Method

M evaluated indicators form M-order comparison matrix.

$$A = \{a_{ij}\}_{M \times N} \approx \{W_i/W_j\}_{M \times N} \tag{1.7}$$

If evaluators estimate are inconsistent with $a_{ij}, a_{ij} \neq W_i/W_j$ that is, $a_{ij}W_j - W_i \neq 0$, we can choose a set of weights $(W_1, W_2, \ldots, W_M)$ to minimize the error of the sum of squares, namely,

$$\min\left\{ Z = \sum_{i=1}^{M} \sum_{j=1}^{M} (a_{ij}W_j - W_i)^2 \right\}$$

$$\sum_{j=1}^{M} W_j = 1 \tag{1.8}$$

1.3.4.2 Eigenvector Method

From (1.7) we know that $AW \approx MW$. So

$$(A - MI)W \approx 0 \tag{1.9}$$

where *I* is the unit matrix, $W = (W_1, W_2, \ldots, W_m)^T$. If an estimate is consistent, the result of (1.9) will be zero. Otherwise, it becomes necessary to solve the following equation:

$$/A - \lambda_\mu I / = 0 \qquad (1.10)$$

where $\lambda\mu$ is the largest characteristic value for the matrix *A*, and the corresponding eigenvector of $\lambda\mu$ is the weight vector.

1.3.4.3 Other Weighting Methods

Many methods can determine weights based on the previously outlined theories. Principal component analysis is introduced below. We can use Equation (1.2) to solve *M* characteristic values if the contribution rate of the first component Y_1

$$a_1 = \frac{\lambda_1}{\sum\limits_{j=1}^{m} \lambda_j} > 0.85 \qquad (1.11)$$

The corresponding feature vector of λ_1 can be viewed in relation to the weight vectors of the original indicators. If a_1 is not large enough, we can take the combination of product of the corresponding features and contribution rates from the previous main components and calculate weights using the normalized weight vector.

1.4 Comparative Evaluation and Logical Framework Approach

Comparative evaluation and the logical framework approach are two methods for evaluating project planning, implementation, assessing effectiveness after completion, and determining impacts. Through objective analysis and evaluation of project activities, an organization can determine whether a project target can be reached or decide whether a project or plan is reasonable and effective. These approaches aim to achieve efficiency through analysis and evaluation, identify lessons learned and the reasons for success or failure, utilize timely and accurate information for future projects, improve decision making and investment management, and evaluate recommendations for improvement.

Project evaluation was introduced in the United States in the mid-19th century. By the 20th century, it was widely accepted by many countries, the World Bank, large Asian banks, and aid organizations worldwide, particularly related to investment activities.

1.4.1 Comparison Method for Main Indicators

Comparison methods assess and compare "before" and "after" indicators. The two comparison methods for main indicators are the *before-and-after comparison* and the *with-and-without comparison* approaches. Comparison methods are effective for determining the value and effectiveness of a project before and after implementation. Feasibility and economic studies are usually conducted in the early stages of a project to predict outcomes. Comparisons can determine whether planning, decision making, quality control, and implementation meet expectations and standards.

The *with-and-without comparison* method compares the results of a project in progress with results that would have been achieved without the project to measure the true impact from implementing the project as shown in Figure 1.6. This technique is often used to evaluate effectiveness and impacts of existing projects and determine whether to proceed with new projects. The advantage is that an organization can control project costs and resources after comparing two types of results. Limiting a comparison only to before and after conditions may distort the real effects of a project. The with-and-without comparison approach requires information about inputs and outputs. Post-project evaluation should eliminate nonproject factors to determine the true results. The steps of two comparative methods are outlined below.

1.4.1.1 Method 1

1. Evaluate the indicators to determine the objective of the comparison.
2. Before a project is implemented, collect time series data and project results after analyzing the data.
3. Using statistical analysis before implementation of the project, predict the value of each index.
4. The difference between predicted and actual results after implementation represents the impact of the project.
5. Investigate outside factors to determine whether they exerted any influence on the project.

This method is suitable for analyzing historical data even if no project is planned. It allows organizations to track rising and falling trends.

1.4.1.2 Method 2

1. Determine the evaluation indicators.
2. Select a group of comparable objects via random sampling to determine test subjects and controls.
3. Measure each set of objects before the project starts.

4. Implement the project using both experimental and control groups.
5. Monitor the experimental and control groups to observe impacts on results.
6. Measure the value of target value after implementation of the project.
7. Compare objects in both groups before and after implementation of the project to determine changes caused by the project.
8. Search for causes other than the differences between two groups. If other factors are revealed, determine their influence on the project.

Method 2 can be used to analyze the impacts of projects on individuals but such specific application requires more time and funding. The method can evaluate the role and impact of each variable. Comparisons are useful for planning and analyzing nonimplementation aspects of projects.

1.4.2 Logical Framework Approach

The logical framework approach (LFA) was developed by the U.S. Agency for International Development in 1970 using design, planning, and evaluation tools. Some aspects of the approach are used by international organizations in project planning, management, and evaluation to aid decision making and analyze key factors and issues.

1.4.2.1 Basic Concept

The logical framework approach is a conceptual method that uses a simple block diagram to analyze complex projects and define meanings and relationships to make a project easier to understand. This approach identifies core issues to be resolved, determines start-up steps, and analyzes impacts. *Problem trees* are used to describe causal relationships related to the problem. The next step is completing an *objective tree* to produce a planning matrix. The logical framework approach for project planners and evaluators is useful for analyzing project objectives and developing ways to achieve the objectives. The core concepts of the approach involve logical causes and effects among project layers. For example, "if" certain internal and external conditions are met, "then" a certain outcome will be produced. The logical framework model is a 4 × 4 matrix. Table 1.6 is an example.

In the course of evaluation of an investment project, a logical framework providing clear descriptions of lenders, borrowers, and beneficiaries will clarify the purpose and content of a project and improve and refine preparation, decision making, and evaluation procedures. The logical framework approach is based on project development and change. Achieving the desired results requires accurate cost effectiveness analysis, multischeme comparisons to achieve desired results at lower cost, and assessment of project management. In applying the logical framework

Table 1.6   Logical Framework Model

Level	*Indicators*	*Authentication Method*	*Important External Conditions*
Objective	Indicators of targets	Detection and monitoring instruments and methods	Main conditions to achieve goals
Purpose	Indicators of purposes	Detection and monitoring instruments and methods	Main conditions to achieve purpose
Output	Quantitative indicators of outputs	Detection and monitoring instruments and methods	Main conditions to achieve outputs
Input	Quantitative indicators of inputs	Detection and monitoring instruments and methods	Main conditions for implementation of investment

planning and evaluation method, one of the main tasks is clearly defining project objectives by utilizing:

1. Clear and measurable objectives
2. Different levels of goals that will be linked ultimately
3. Main content
4. Measures of success
5. Planning and design of main assumptions
6. Monitoring progress of approach
7. Implementation using required resources and inputs

1.4.2.2 Goal Levels

In the logical framework approach, the goals and causal relationships are classified into four levels: objectives, purposes, outputs, and inputs.

1. Objectives are the requirements to achieve the ultimate goal of a project. The achievement of objectives can be measured by several indicators. The SMART (specific, measurable, action-oriented, realistic, and time-related) principles should be applied when the objectives are determined.
2. Purposes indicate why a project should be implemented (direct effect and role). Social and economic benefits for target groups should be considered.

Purposes are usually determined by project builders or independent evaluation agencies.

3. Outputs indicate *what a project did*—direct results that may be measured.
4. Inputs are financial and time resources invested to implement a project.

The levels are connected from the bottom up. The bottom level ensures that certain resources will be invested and well managed, and will produce the expected outputs. The next level covers project outputs and changes in social or economic relationships. The purposes level relates to expected outcomes.

1.4.2.3 Vertical Logic of Causality

In the logical framework approach, *vertical logic* can be used to elaborate the content and objectives of each level and the causal relationships between the upper and lower levels. Figure 1.5 shows the logical framework of the four levels and important restrictions based on assumptions about external conditions and risks. Important assumptions are possible outcomes. Note that a project manager has no control over external conditions, risks, or restrictions. Examples of factors that cannot be controlled are (1) changes of the natural project site environment; (2) serious repercussions due to changes of government policy, planning, development strategies, and other major impacts; and (3) management problems that isolate the project inputs and outputs.

Project assumptions are important factors and should be based on analysis of pre-project conditions, input and output targets, macro objectives, and levels of uncertainties. Project success or failure depends on the factors on which assumptions are based; thus the assumptions must be analyzed carefully.

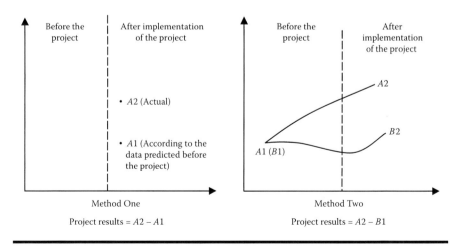

Figure 1.5   Vertical logic of causality.

1.4.2.4 Horizontal Logic

Vertical logic distinguishes logical frameworks to evaluate the project hierarchy but is insufficient for analyzing project implementation requirements. Horizontal logic analysis is used to measure project resources and results based on main verification indicators. The four goal levels indicate a logical 4 × 4 framework. Table 1.7 shows the relationship of verification and validation methods and content.

1.4.2.5 Verifying Indicators

The objective and measurable verification indicators of different goals of the logical framework model should be determined to illustrate the results at every level. To verify the degree of success, the verification indicators of the logical framework should have the following characteristics:

1. Clear quantitative indicators for measuring success.
2. Highlighting key indicators to illustrate the need for the project.
3. Clear verification of relationships between indicators and corresponding objectives.
4. Verification of correspondence between indicators and objectives.
5. Verification that indicators are complete, adequate, and accurate.
6. Verification that indicators are objective, concrete, and not subject to outside changes.

Table 1.7   Horizontal Logic

Goal Level	Validation Index	Authentication Sources and Methods
Impact/objective	Degree of influence	Sources: documents, official statistics, project beneficiaries
		Methods: data analysis, investigation
Role/purpose	Size of action	Sources: beneficiaries
		Methods investigation
Outputs	Qualitative and quantitative outputs at different stages	Sources: Project records, reports, beneficiaries
		Methods: data analysis, investigation
Inputs	Resource costs, quantities, locations, time requirements	Sources: project evaluation report, plans, investor agreements, other documents

7. Indirect indicators. It may be difficult to verify project indicators directly. The relationship between indirect indicators and verification objects must be clear.
8. Verification of accuracy of indicators. The indicators should include clear definitions based on quantitative and qualitative data.

1.4.2.6 Verification Method

Verification methods are based on the horizontal logic indicators from the logical framework model. Verification methods are based on data type and collection methods and information sources. More specifically:

1. Data types should be consistent with indicator requirements. Each indicator level has different data requirements; data requirements must be specific.
2. The reliability of information sources must be demonstrated. Sources that can save time and money should be selected first. Information may be obtained from project staff, local authorities, and official documents.
3. Data collection: After the types and sources of data are determined, the investigation forms can be designed. Management may need to set standards to ensure data quality. If a sampling method is used, sample size, content, and statistical standards should be defined.

The logical framework approach is a procedure that aids thinking. It focuses on important factors such as the *who*, *what*, *when*, *why*, and *how*. Although preparing to utilize the logical framework technique is difficult and time-consuming, the method allows decision makers, managers, and evaluators to review project goals, improve project management, and make effective decisions.

1.4.2.7 Logical Framework Approach Used in Postevaluation

Project evaluation solves three issues: (1) adjustment of the original goals and objectives of the project; (2) determining whether the benefits of the project were achieved and at what level; and (3) assessing risks incurred and their potential to affect further projects. A project evaluation must determine (1) whether the original goals and objectives were achieved and whether they were reasonable; (2) whether the benefits were received and what lessons were learned; and (3) whether the project is sustainable. The logical framework can be applied to project evaluation and also to postassessments. The basic format is shown in Table 1.8.

Although the logical framework approach has many advantages for postevaluation of projects, it also presents some limitations:

1. Early in a project, stress and external factors may lead to rigid management.
2. The logical framework approach is effective only for the general analysis of policy issues such as income distribution, employment opportunities,

Table 1.8 Logical Framework for Postevaluation

Goal Level	Verification and Contrast Indicators			Reasons		
	Original Target Item	Indicators of Implementation	Differences or Changes	Main Internal Factors	Main External Factors	Sustainability (Risk)
Objective (impact)						
Purpose (role)						
Outputs (results)						
Inputs (conditions)						

resource access, local participation, cost and strategic factors, feasibility, and relationships of external conditions.
3. The logical framework approach is a mode of thinking. It cannot replace benefit analysis, scheduling, economic and financial analysis, cost–benefit analysis, environmental impact assessment, and other specific techniques.

1.5 Analytic Hierarchy Process

AHP is a decision method developed in the 1970s by Dr. Thomas L. Saaty, a U.S. operations researcher. It is a flexible and systematic decision-making tool for analyzing complex, unstructured multiobjective, multicriteria, multielement, multilevel problems that incorporate qualitative and quantitative issues. AHP techniques are classified as complementary and reciprocal. This section discusses the reciprocal model. The four steps of modeling are:

1. Hierarchical structuring
2. Construction of judgment matrix
3. Single level ordering and consistency check
4. Total ordering and consistency check

1.5.1 Hierarchical Structuring

For some problems that need evaluation, it is important to take a holistic approach and obtain a clear picture of the relationships of systems and their

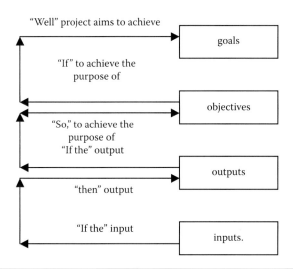

Figure 1.6 Methods with and without contrast.

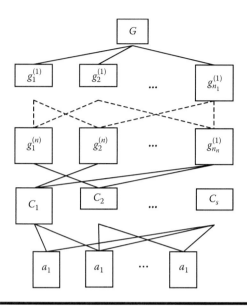

Figure 1.7 Hierarchical structure model.

environments, contained subsystems, molecular systems, and elements. It is also critical to understand the relationships of subsystems at the same level and elements of a system, subsystems and elements at different levels and merge common elements into a group as a hierarchy of one model containing three levels (Figure 1.7).

1. The highest (general) goal level is only one element of decision analysis (layer G in the figure).
2. The middle (or target) may require more than one level to accommodate all subgoals included in the general goal. Various elements such as constraints, guidelines, and strategies are included in layers G and C in the figure.
3. The lowest (alternatives) level indicates all possible options for reaching the goal (layer A in the figure).

1.5.2 Constructing Judgment Matrix

Assume that M elements are relatively important for the upper element. Elements i and j (both greater than 0 and smaller than m) are compared, and a_{ij} denotes relative importance. The final result is called the comparison matrix. Table 1.9 shows relative importance on a 1 to 9 scale based on a common scale partitioning method. Of course, for specific applications, we can adjust the relative importance scale as needed, for example, we can use a 1 to 3 or 1 to 0 scale or even an exponential scale.

Table 1.9 Importance Matrix

Scale	Definition	Meaning
1	Equally important	Two elements are equally important
3	Slightly more important	One element is slightly more important
5	Obviously more important	One element is obviously more important
7	Strongly more important	One element is strongly more important
9	Extremely more important	One element is extremely more important
2, 4, 6, 8	Middle value of adjacent scale	Middle is comparatively important
Scale reciprocal	Compare converse	The importance of *i* and *j* is a_{ij}

If judgment matrix $A = (a_{ij})_{m \times m}$ meets the following conditions

$$\text{(a) } a_{ij} > 0; \ a_{ij} > 0; \ i,j = 1,2,\ldots,m$$

$$\text{(b) } a_{ij} = 1, i = 1,2,\ldots,m; \tag{1.12}$$

$$\text{(c) } a_{ij} = \frac{1}{a_{ij}}, i, j, \cdots, m$$

we call $A = (a_{ij})_{m \times m}$ a positive reciprocal matrix.

1.5.3 Single Level Ordering and Consistency Check

For the standard C, m elements $a_1, a_2, \ldots, a_m$ are compared according to the previously outlined rules to yield the judgment matrix shown in Table 1.10.

If judgment matrix $A = (a_{ij})$ is a positive reciprocal matrix, the characteristic equation is

$$| AW - \lambda_{\max} W | = 0 \tag{1.13}$$

Table 1.10 Single-Level-Order Judgment Matrix

C	a_1	a_2	$\cdots$	a_m
a_1	1	a_{12}	$\cdots$	a_{1m}
a_2	a_{21}	1	$\cdots$	a_{2m}
$\vdots$	$\cdots$	$\cdots$	$\ddots$	$\cdots$
a_m	a_{m1}	a_{m2}	$\cdots$	1

We can determine the maximum value λ_{max} of the judgment matrix A and corresponding eigenvector W; normalizing W we can calculate the weight vector of m elements $a_1, a_2, \ldots, a_m$:

$$W = (w_1, w_2, \cdots, w_m)^T$$

If elements of a positive reciprocal matrix have transitive property:

$$a_{ik} = a_{ij} \cdot a_{jk}, i, j, k = 1, 2, \cdots, m \tag{1.14}$$

we say the matrix meets consistency conditions. Objective analysis is complicated and very few judgment matrices meet the consistency condition. Only when a judgment matrix meets the consistency condition will it have the unique nonzero maximum value; the rest of the characteristics are valued at zero. The judgment matrix A's maximum value $\lambda_{max} = m$.

Because the judgment matrix is a positive reciprocal matrix, it does not meet consistency conditions. Matrix A has the condition $\lambda_{max} \geq m$. Not every eigenvalue equals zero. If a judgment matrix mainly meets the consistency condition, then λ_{max} is slightly greater than m and the other eigenvalues are approximately equal. Only when judgment matrix meets the consistency conditions are the W and weight vector meaningful. A check is needed to ensure that a judgment matrix achieves satisfactory consistency. Set the value of a judgment matrix A as

$$\lambda_1 = \lambda_{max}, \lambda_2, \ldots, \lambda_m$$

Because A is a positive reciprocal matrix, then

$$\text{tr}(A) = \sum_{i=1}^{m} a_{ii} = m$$

The sum of all of A's eigenvalues is

$$\lambda_{\max} + \sum_{i=1}^{m} \lambda_i = m = tr(A)$$

Then

$$\left| \sum_{i=2}^{m} \lambda_i \right| = \lambda_{\max} - m$$

The consistency index

$$C \cdot I = \frac{\lambda_{\max} - m}{m - 1} \tag{1.15}$$

where m is the order and $\lambda_{\max}$ is the maximum eigenvalue of the judgment matrix. Generally, the larger the $C \cdot I$, the more consistency deviates and vice versa. For ease in judging, we use an average random consistency index $R \cdot I$. Table 1.11 shows how $R \cdot I$ changes with order.

The consistency index $C \cdot I$ divided by the average random consistency index $R \cdot I$ yields the consistency ratio designated $C.R$:

$$C \cdot R = \frac{C \cdot I}{R \cdot I} \tag{1.16}$$

$C \cdot R$ is used to judge the consistency of a matrix. When $C \cdot R$ is less than 0.1, $W = (w_1, w_2, \ldots w_m)^T$ is acceptable. Otherwise the matrix must be adjusted until $C.R$ is less than 0.1.

1.5.4 Hierarchy Total Ordering and Consistency Check

We already set the weight vectors of elements. To move the elements of all the layers of the weight vectors to the overall goal, we must compute the final weight from top to bottom. This section outlines the process.

The system has s layers, and the element number of the k-th layer is n_k, $k = 1, 2, \ldots, s$. The first layer is the unique (overall) goal, $n_1 = 1$. The weight vector of the second layer's n_2 elements is

$$W^{(2)} = \left(w_1^{(2)}, w_2^{(2)}, \cdots, w_{n2}^{(2)} \right)^T$$

Table 1.11 R·I

Order	1	2	3	4	5	6	7	8	9	10	11	12	13	14	15
RI	0	0	0.52	0.89	1.12	1.26	1.36	1.41	1.46	1.49	1.52	1.54	1.56	1.58	1.59

Set the $k-1$ layer's weight vector to the goal as

$$W^{(k-1)} = \left(w_1^{(k-1)}, w_2^{(k-1)}, \cdots, w_{n_{k-1}}^{(k-1)}\right)^T$$

The weight vector of the n_k elements on layer k with regard to element j of the $k-1$ layer is

$$P_j^{(k)} = \left(p_{1j}^{(k)}, p_{2j}^{(k)}, \cdots, p_{nkj}^{(k)}\right)^T$$

If uncorrelated, the value is 0. Then the weight matrix of k layer's elements to $k-1$ layer's n_{k-1} elements is

$$P^{(k)} = (p_1^{(k)}, p_2^{(k)}, \cdots, p_{n_{k-1}}^{(k)})^T$$

$$= \begin{bmatrix} p_{11}^{(k)} & p_{12}^{(k)} & \cdots & p_{1n_{k-1}}^{(k)} \\ p_{21}^{(k)} & p_{22}^{(k)} & \cdots & p_{2n_{k-1}}^{(k)} \\ \cdots & \cdots & \cdots & \cdots \\ p_{n_k 1}^{(k)} & p_{n_k 2}^{(k)} & \cdots & p_{n_k n_{k-1}}^{(k)} \end{bmatrix}$$

The weight vector of the n_k elements on layer k with regard to the general goal G is the product of

$$P^{(k)} = \left(p_1^{(k)}, p_2^{(k)}, \cdots, p_{n_{k-1}}^{(k)}\right)^T \text{ and } W^{(k-1)} = \left(w_1^{(k-1)}, w_2^{(k-1)}, \cdots, w_{n_{k-1}}^{(k-1)}\right)^T$$

$$W^{(k)} = \left(w_1^{(k)}, w_2^{(k)}, \cdots, w_{n_k}^{(k)}\right)^T = P^{(k)} \cdot W^{(k-1)}$$

$$= \begin{bmatrix} p_{11}^{(k)} & p_{12}^{(k)} & \cdots & p_{1n_{k-1}}^{(k)} \\ p_{21}^{(k)} & p_{22}^{(k)} & \cdots & p_{2n_{k-1}}^{(k)} \\ \cdots & \cdots & \cdots & \cdots \\ p_{n_k 1}^{(k)} & p_{n_k 2}^{(k)} & \cdots & p_{n_k n_{k-1}}^{(k)} \end{bmatrix} \begin{pmatrix} w_1^{(k-1)} \\ w_2^{(k-1)} \\ \vdots \\ w_{n_{k-1}}^{(k-1)} \end{pmatrix} \tag{1.17}$$

$$w_i^{(k)} = P_i^{(k)} \cdot W^{(k-1)} = \left(p_{i1}^{(k)}, p_{i2}^{(k)}, \cdots, p_{in_{k-1}}^{(k)}\right) \cdot \left(w_1^{(k-1)}, w_2^{(k-1)}, \cdots, w_{n_{k-1}}^{(k-1)}\right)^T$$

$$= \sum_{j=1}^{n_{k-1}} p_{ij}^{(k)} w_j^{(k-1)}, i = 1, 2, \cdots, n_k .$$

Equation (1.17) is a recurrence formula that can calculate the weights of the n_k elements on layer k by the weight of the n_{k-1} elements on layer $k-1$. From Equation

(1.17), the weight vectors of the n_s elements on layer s with regard to the general goal on the first layer can be calculated by Equation (1.18):

$$W^{(s)} = P^{(s)}W^{(s-1)} = P^{(s)}P^{(s-1)}W^{(s-2)} = P^{(s)}P^{(s-1)}P^{(s-2)} \cdots P^{(3)}W^{(2)} \quad (1.18)$$

The steps for calculating the multilayer hierarchical weight vector are:

Step 1: Define the structure of the system and the number of elements on each layer. Assume there are s layers and the number of elements on layer k is:

$$n_k, k = 1, 2, \cdots, s$$

Step 2: The weight vectors of the n_s elements on the second layer to the general goal of the first layer:

$$W^{(2)} = (w_1^{(2)}, w_2^{(2)}, \cdots, w_{n_2}^{(2)})^T$$

Step 3: Calculate the weight matrix of the n_k elements on layer k to the n_{k-1} elements on layer $k - 1$:

$$P^{(k)}, k = 3, 4, \cdots, s$$

Step 4: Calculate the weight vector $W^{(s)}$ of the n_s elements on layer s with regard to the general goal on the first layer by using Equation (1.18).

We must do a consistency check for multilayer hierarchical weight vectors. Good consistencies at individual levels may still allow inconsistencies to appear in the overall matrix. We must ensure that the synthetic weight vectors and the calculation process are the same as calculating the weight vector of hierarchical structure—from high to low level. Assume that the consistency index, random consistency index, and consistency ratio of the n_k elements on layer k with regard to element j of layer $k - 1$ are:

$$CI_j^{(k)}, RI_j^{(k)}, CR_j^{(k)}, j = 1, 2, \cdots, n_{k-1}$$

The weight vector of the n_{k-1} elements on layer $k - 1$ with regard to the general goal on the first layer is:

$$W^{(k-1)} = \left(w_1^{(k-1)}, w_2^{(k-1)}, \cdots, w_{n_{k-1}}^{(k-1)} \right)^T$$

Equations (1.19), (1.20), and (1.21) show the calculation process:

$$CI^{(k)} = \left(CI_1^{(k)}, CI_2^{(k)}, \cdots, CI_{n_{k-1}}^{(k)} \right) W^{(k-1)} \quad (1.19)$$

$$RI^{(k)} = \left(CI_1^{(k)}, CI_2^{(k)}, \cdots, CI_{n_{k-1}}^{(k)}\right) W^{(k-1)} \tag{1.20}$$

$$CR^{(k)} = \frac{CI^{(k)}}{RI^{(k)}}, k = 3,4,\cdots,s \tag{1.21}$$

If $CR^{(k)} < 0.1$, the judgment matrix from the first layer to the k layer satisfies the consistency check on the whole and the comprehensive weight vector $W^{(k)} = (w_1^{(k)}, w_2^{(k)}, \cdots, w_{n_k}^{(k)})^T$ is acceptable. It is difficult to adjust if not satisfy the overall consistency check. Overall consistency is not emphasized in practical applications.

1.6 DEA Relative Efficiency Evaluation

The data envelopment analysis (DEA) model is mainly used to compare multiple inputs and outputs of groups and production functions. The method was created by American operations research experts Charnes, Cooper, and Rhodes in 1978. Its basic idea is to treat every evaluation unit as a decision-making unit (DMU). Numerous DMUs constitute a decision-making group. Efficient production is determined by calculating input:output ratios of DMUs and using weights of input and output indicators as variables. DMUs are classified as DEA effective, weakly DEA effective, and non-DEA effective to indicate improvement paths. In recent years, DEA has been widely used in economic and social management projects. This section introduces the relative efficiency evaluation method of DEA and uses the C²R model as an example.

1.6.1 DEA Efficiency Evaluation Index and C²R Model

1.6.1.1 DEA Efficiency Evaluation Index

Assume there are n decision-making units, and each unit has m inputs and p outputs. Figure 1.8 shows the system. x_{ij} is the i input of the decision-making unit $j, x_{ij} > 0; y_{rj}$ indicates the quantity of output r of the decision-making unit $j, y_{rj} > 0$; v_i indicates the weight of the input $I, v_i \geq 0$; and u_r is the weight of output r, $u_r \geq 0$ $(i = 1,2,\ldots,m; j = 1,2,\ldots,n; r = 1,2,\ldots,p)$. x_{ij} and y_{rj} are known numbers that can be calculated from historical data and statistics. The weight of u_r and v_i can be determined through the model.

For the decision-making unit j $(j = (1,2,\ldots,n))$, the efficiency evaluation index is

$$h_j = \frac{\sum\limits_{r=1}^{p} u_r y_{rj}}{\sum\limits_{i=1}^{m} v_i x_{ij}} \tag{1.22}$$

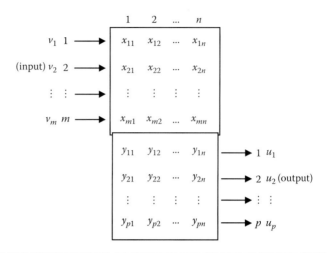

Figure 1.8   Multiple input and multiple output evaluation.

Appropriate weight vectors u and v yield $h_j \leq 1$. Generally speaking, the bigger h_j is, the smaller the input will be.

1.6.1.2 C²R Design

If we treat the maximization of efficiency evaluation index of the DMU j_0 as an objective and the efficiency evaluation index $h_j \leq 1 (j = 1,2,\ldots,n)$ of all the DMUs in the decision group as a constraint condition, the model will be

$$
\begin{cases}
\max h_0 = \dfrac{\displaystyle\sum_{r=1}^{p} u_r y_{rj_0}}{\displaystyle\sum_{i=1}^{m} v_i x_{ij_0}} \\[2em]
s.t. \dfrac{\displaystyle\sum_{r=1}^{p} u_r y_{rj}}{\displaystyle\sum_{i=1}^{m} v_i x_{ij}} \leq 1, (1 \leq j \leq n) \\[2em]
v \geq 0, u \geq 0
\end{cases}
\tag{1.23}
$$

This equation is the fundamental DEA C²R model. Assume that the input and output vectors of the j_0 DMU are, respectively:

$$x_0 = (x_{1j_0}, x_{2j_0}, \cdots, x_{mj_0})^T$$

$$y_0 = (y_{1j_0}, y_{2j_0}, \cdots, y_{pj_0})^T$$

The input and output vectors of the j DMU are, respectively:

$$x_j = (x_{1j}, x_{2j}, \dots, x_{mj})^T$$

$$y_j = (y_{1j}, y_{2j}, \dots, y_{pj})^T$$

The weight vectors of the input and output indices are, respectively:

$$v = (v_1, v_2, \cdots, v_m)^T$$

$$u = (u_1, u_2, \cdots, u_p)^T$$

The C²R model [Equation 1.23] can be expressed as a matrix:

$$(\bar{p}) \begin{cases} \max h_0 = \dfrac{u^T y_0}{v^T x_0} \\[2mm] s.t. \dfrac{u^T y_j}{v^T x_j} \le 1, (1 \le j \le n) \\[2mm] v \ge 0, u \ge 0 \end{cases} \tag{1.24}$$

$t = 1/v^T x_0$, $\omega = tv$, $\mu = tu$ and then $\bar{p}$ can be converted to a linear programming problem.

$$(p) \begin{cases} \max V_p = \mu^T y_0 \\[2mm] s.t. \omega^T x_j - \mu^T y_j \ge 0, (1 \le j \le n) \\[2mm] \omega^T x_0 = 1 \\[2mm] \omega \ge 0, \mu \ge 0 \end{cases} \tag{1.25}$$

The dual problem of (p) is

$$(D)\begin{cases} \min V_D = \theta \\[2mm] \sum_{j=1}^{n} x_j \lambda_j + s^- = \theta x_0 \\[2mm] \sum_{j=1}^{n} y_j \lambda_j - s^+ = y_0 \\[2mm] \lambda_j \geq 0, (1 \leq i \leq n) \\[2mm] s^+ \geq 0, s^- \geq 0 \end{cases} \qquad (1.26)$$

and $S^- = \left(S_1^-, S_2^- \cdots S_m^-\right)^T$ and $S^+ = \left(S_1^+, S_2^+ \cdots S_m^+\right)^T$ are slack variables; (p) and (D) are both expressions of the model.

1.6.2 DEA Validity Judgment

Using model (p), the validity of the DMU j_0 can be determined as follows:

1. If the optimal solution of linear programming (p) is ω^0, μ^0 and $V_p = (\mu^0)^T y_0 = 1$, j_0 is weakly DEA effective.
2. If the optimal solution of (p) is ω^0, μ^0 and $V_p = (\mu^0)^T y_0 = 1, \omega^0 > 0, \mu^0 > 0, j_0$ is DEA effective.
3. If the optimal solution of (p) is ω^0, μ^0 and $V_p = (\mu^0)^T y_0 < 1, j_0$ is non-DEA effective.

Using model (D), DEA validity of the DMU j_0 can be determined:

1. If the optimal value $V_D = 1, j_0$ is weakly DEA effective.
2. If the optimal value $V_D = 1$ and every optimal solution $\lambda^0 = (\lambda_1^0, \lambda_2^0, \cdots \lambda_n^0)^T$, s^{0-}, s^{0+}, θ^0 satisfies the condition $s^{0-} = 0$, $s^{0+} = 0$, j_0 is DEA effective.
3. If the optimal value $V_D < 1, j_0$ is non-DEA effective.

Obviously, If j_0 is DEA effective, it is also weakly DEA effective. If a DMU is weakly DEA effective, the efficiency evaluation index of the DMU can achieve the optimal value of 1 and is technology effective. When a DMU is DEA effective, it is technology effective and scale effective. A DMU that is non-DEA effective is neither technology effective nor scale effective. Charnes, Cooper, and Rhodes cited the concept of non-Archimedean infinitesimal. Thus, weight vectors can be required to exceed 0, and the C^2R model can be solved with simplex methods to judge DEA validity. Let ε represent a non–Archimedean infinitesimal; then the C^2R model can be

$$(p_\varepsilon)\begin{cases}\max \mu^T y_0 = V_p \\ s.t.\omega^T x_j - \mu^T y_j \geq 0,(1\leq j\leq n) \\ \omega^T x_0 = 1 \\ \omega^T \geq \varepsilon \hat{e}^T,\mu \geq \varepsilon e^T\end{cases} \quad (1.27)$$

Where $\hat{e}^T = (1,1,\cdots,1)$ is a vector of m dimension and $e^T = (1,1,\ldots,1)$ is a vector of p dimension. The dual program of (p_ε) is

$$(D_\varepsilon)\begin{cases}\min[\theta - \varepsilon(\hat{e}^T s^- + e^T s^+)] = V_{D_\varepsilon} \\ s.t. \sum_{j=1}^{n} x_j\lambda_j + s^- = \theta x_0 \\ \sum_{j=1}^{n} y_j\lambda_j - s^+ = y_0 \\ \lambda_j \geq 0,(1\leq i\leq n) \\ s^+ \geq 0, s^- \geq 0\end{cases} \quad (1.28)$$

The optimal solution of $(D\varepsilon)$ is $\lambda^0, s^{0-}, s^{0+}, \theta^0$ and the method to judge the validity of a DMU with the C^2R model which with the non-Archimedean infinitesimal ε is:

1. If $\theta^0 = 1$, the DMU j_0 is weakly DEA effective.
2. If $\theta^0 = 1$ and $s^{0-} = 0$, $s^{0+} = 0$, the DMU j_0 is DEA effective.
3. If $\theta^0 < 1$, the DMU j_0 is non-DEA effective.

1.6.3 Improving DEA Validity

Improving DEA validity requires adjustment of inputs and outputs of a DMU to transform it from non-DEA effective to DEA effective. The simplest improvement method using the C^2R model with the non-Archimedean infinitesimal ε is shown in Equation (1.29). Assume $\lambda^0, s^{0-}, s^{0+}, \theta^0$ is the optimal solution of $(D\varepsilon)$ and let

$$x_0 = \theta^0 x_0 - s^{0-}, \hat{y}_0 = \theta^0 x_0 + s^{0+} \quad (1.29)$$

$(\hat{x}_0, \hat{y}_0)$ constitute a new decision-making unit. It is easy to prove that $(\hat{x}_0, \hat{y}_0)$ is effective compared with the original unit. Denote $\Delta x_0 = x_0 - \hat{x}_0$ and $\Delta y_0 = \hat{y}_0 - y_0$. Δx_0 represents input surplus; Δy_0 is an output deficit. A DMU of non-DEA effective j_0, can become DEA effective by eliminating the input surplus to offset the output deficit.

1.7 Chapter Summary

As systems science progresses rapidly, systems evaluation and research attract increasing attention. System evaluation methods and their models serve as the core of this chapter. Group techniques, Delphi, brainstorming, and other common evaluation methods in which information is obtained via a series of steps are strongly subjective. The analysis of a problem is based on the construction of an evaluation index system. To develop an index system and understand its statistics, methods, and weights, we must study the inherent characteristics of the index by quantitative analysis and examine the information behind the data.

AHP and DEA are often applied to complex system evaluation and analysis. AHP can effectively break down a problem into levels and elements. Comparisons of elements reveal different options and weight and provide a reliable basis for evaluation. DEA is a multiobjective evaluation method suitable for evaluating the relative efficiencies of DMUs containing multiple inputs and outputs.

Chapter 2

Grey System Evaluation Models

2.1 Introduction

Several new grey system evaluation models will be introduced in this chapter, including the generalized grey incidence model, grey incidence models based on visual angles of similarity and nearness, grey cluster evaluations based on end-point and center-point triangular whitenization functions, and multiattribute grey target decision models. The modeling software can be downloaded at no charge from the Institute for Grey Systems Studies at the Nanjing University of Aeronautics and Astronautics (http://igss.nuaa.edu.cn/).

The new generalized grey incidence model is based on an overall or global perspective that contains an absolute degree of grey incidence model as its main body, the relative degree of grey incidence model based on initial value transformation and absolute degree of grey incidence model, and the synthetic degree of grey incidence model composed of the absolute degree of grey incidence model and the relative degree of grey incidence model. Grey incidence models based on similarity nearness measure the geometric similarities of the shapes of sequences X_i and X_j. The more similar the geometric shapes of X_i and X_j, the greater similitude degree of grey incidence and vice versa. When the concept of nearness is employed to measure the spatial nearness of sequences X_i and X_j, the closer X_i and X_j, the greater the nearness degree of grey incidence and vice versa. This concept is meaningful only when the sequences X_i and X_j possess similar meanings and identical units. Otherwise, it does not have practical significance.

The two types grey cluster evaluations based on triangular whitenization functions are (1) the evaluation based on end-point triangular whitenization weight functions and (2) the evaluation based on center-point triangular whitenization weight functions. The grey cluster evaluation model based on end-point triangular whitenization weight functions is applicable when the grey boundary is clear and the most possible point of each likely grey class is unknown; the grey cluster evaluation based on center-point triangular whitenization weight functions is applicable when the most possible point of each grey class is known but their boundaries are unclear. These two assessment models are based on moderate measures of triangular whitenization weight functions.

The critical value of a grey target is the dividing point between positive and negative, that is, the zero point of a multiattribute grey target decision model, that is, it determines whether an objective "hits the bull's eye." The different uniform effects measure functions based on different decision objectives related to benefits, costs, and other factors. Accordingly, the decision objectives involving different meanings, dimensions, or natures may be transformed into a uniform effect measure. The matrix of synthetic effect measures can be obtained easily.

2.2 Generalized Grey Incidences Model

2.2.1 Absolute Grey Incidence Model

Definition 2.1

Let $X_i = (x_i(1), x_i(2), \ldots, x_i(n))$ be the data sequence of a system's behavior, D, a sequence operator satisfying that

$$X_i D = (x_i(1)d, x_i(2)d, \ldots, x_i(n)d)$$

Where $x_i(k)d = x_i(k) - x_i(1)$, $k = 1, 2, \ldots, n$, then D is the zero starting point operator with $X_i D$ as the image of X_i, denoted by

$$X_i D = X_i^0 = (x_i^0(1), x_i^0(2), \cdots, x_i^0(n))$$

The corresponding zigzagged line of the sequence X_i, X_j, X_i^0, X_j^0 is also recorded as

$$X_i, X_j, X_i^0, X_j^0$$

Let

$$s_i = \int_1^n (X_i - x_i(1)) \, dt = \int_1^n X_i^0 \, dt \tag{2.1}$$

$$s_i - s_j = \int_1^n (X_i^0 - X_j^0)dt \qquad (2.2)$$

$$S_i - S_j = \int_1^n (X_i - X_j)dt \qquad (2.3)$$

■

Definition 2.2

The sum of time intervals between consecutive observations of the sequence X_i is called the length of X_i. Note that two one-time intervals may not necessarily generate the same number of observations. For example:

$$X_1 = (x_1(1), x_1(3), x_1(6))$$

$$X_2 = (x_2(1), x_2(3), x_2(5), x_2(6))$$

$$X_3 = (x_3(1), x_3(2), x_3(3), x_3(4), x_3(5), x_3(6))$$

The length of X_1, X_2, X_3 is 5, but the number of observations to each sequence is different. ■

Definition 2.3

Assume that two sequences X_i and X_j are of the same length and $s_i, s_j, s_i - s_j$ are defined as in Equations (2.1) and (2.2). Then

$$\varepsilon_{ij} = \frac{1 + |s_i| + |s_j|}{1 + |s_i| + |s_j| + |s_i - s_j|} \qquad (2.4)$$

is called the absolute degree of grey incidence of X_i and X_j or absolute degree of incidence which measures the similarity of X_i and X_j. The greater the degree of geometric similarity of X_i and X_j, the smaller $|s_i - s_j|$, and the larger ε_{ij}. ■

Proposition 2.1

Assume that X_i and X_j are two one-time interval sequences of the same length, and

$$X_i^0 = (x_i^0(1), x_i^0(2), \cdots, x_i^0(n))$$

$$X_j^0 = (x_j^0(1), x_j^0(2), \cdots, x_j^0(n))$$

are zero images of X_i and X_j. Then

$$|s_i| = \left| \sum_{k=2}^{n-1} x_i^0(k) + \frac{1}{2} x_i^0(n) \right| \qquad (2.5)$$

$$|s_j| = \left| \sum_{k=2}^{n-1} x_j^0(k) + \frac{1}{2} x_j^0(n) \right| \qquad (2.6)$$

$$|s_i - s_j| = \left| \sum_{k=2}^{n-1} (x_i^0(k) - x_j^0(k)) + \frac{1}{2} (x_i^0(n) - x_j^0(n)) \right| \qquad (2.7)$$

From (2.5), (2.6), and (2.7), the value of equation (2.4) can be easily calculated. ■

Proposition 2.2

The absolute degree ε_{ij} of grey incidences satisfies the following conditions:

(i) $0 < \varepsilon_{ij} \le 1$;
(ii) ε_{ij} is related only to the geometric shapes of X_i and X_j and has nothing to do with the spatial positions of X_i and X_j.
(iii) The more X_i and X_j are geometrically similar, the greater ε_{ij};
(iv) When X_i and X_j are parallel, or X_j^0 vibrates around X_i^0 with the area of the parts with X_j^0 on top of X_i^0 equal to the parts with X_j^0 beneath X_i^0, $\varepsilon_{ij} = 1$;
(v) $\varepsilon_{00} = \varepsilon_{ij} = 1$;
(vi) $\varepsilon_{ij} = \varepsilon_{ij}$.

Example 2.1

Assume the sequences:

$$X_0 = (x_0(1), x_0(2), x_0(3), x_0(4), x_0(5), x_0(7)) = (10,9,15,14,14,16)$$

$$X_1 = (x_1(1), x_1(3), x_1(5), x_1(7)) = (46,70,84,98)$$

Compute the absolute degree of incidence ε_{01}.

Solution

(1) Turn X_1 into a sequence with the same corresponding time intervals as X_0 and let

$$x_1(2) = \frac{1}{2}(x_1(1) + x_1(3)) = \frac{1}{2}(46 + 70) = 58$$

$$x_1(4) = \frac{1}{2}(x_1(3) + x_1(5)) = \frac{1}{2}(70 + 84) = 77$$

Then

$$X_1 = (x1(1), x_1(2), x_1(3), x_1(4), x_1(5), x_1(7)) = (46, 58, 70, 77, 84, 98)$$

(2) Turn X_0, X_1 into equal time interval sequences and make

$$x_0(6) = \frac{1}{2}(x_0(5) + x_0(7)) = \frac{1}{2}(14 + 16) = 15$$

$$x_1(6) = \frac{1}{2}(x_1(5) + x_1(7)) = \frac{1}{2}(84 + 98) = 91$$

$$X_0 = (x_0(1), x_0(2), x_0(3), x_0(4), x_0(5), x_0(6), x_0(7)) = (10, 9, 15, 14, 14, 15, 16)$$

$$X_1 = (x_1(1), x_1(2), x_1(3), x_1(4), x_1(5), x_1(6), x_1(7)) = (46, 58, 70, 77, 84, 91, 98)$$

are all one-time interval sequences.

(3) Find the images of zero starting point:

$$X_0^0 = (x_0^0(1), x_0^0(2), x_0^0(3), x_0^0(4), x_0^0(5), x_0^0(6), , x_0^0(7)) = (0, -1, 5, 4, 4, 5, 6)$$

$$X_1^0 = (x_1^0(1), x_1^0(2), x_1^0(3), x_1^0(4), x_1^0(5), x_1^0(6), , x_1^0(7)) = (0, 12, 24, 31, 38, 45, 52)$$

(4) Compute $|s_0|, |s_1|, |s_1 - s_0|$

$$|s_0| = \left| \sum_{k=2}^{6} x_0^0(k) + \frac{1}{2} x_0^0(7) \right| = 20$$

$$|s_1| = \left| \sum_{k=2}^{6} x_1^0(k) + \frac{1}{2} x_1^0(7) \right| = 176$$

(5) Calculate the absolute degree of incidence.

$$\varepsilon_{01} = \frac{1+|s_0|+|s_1|}{1+|s_0|+|s_1|+|s_1-s_0|} = \frac{197}{353} \approx 0.5581$$

2.2.2 Relative Grey Incidence Model

Definition 2.4

Assume $X_i = (x_i(1), x_i(2), \ldots, x_i(n))$ is a behavioral sequence of factor X_i, and D_1 is a sequence operator,

$$X_i D_i = (x_i(1)d_1, x_i(2)d_1, \cdots, x_i(n)d_1)$$

where $\quad x_i(k)d_1 = x_i(k) / x_i(1); \; x_i(1) \neq 0; \; k = 1, 2, \ldots n$ \hfill (2.8)

D_1 is the initial operator, $X_i D_1$ is the image of X_i under the initial operator D_1. $X_i D_1$ is usually denoted by X_i'. ■

Definition 2.5

Assume that X_i and X_j are two sequences of the same length with nonzero initial values, X_i', X_j' are the initial images of X_i, X_j, respectively. Then the absolute degree of grey incidence ε_{ij}' of X_i' and X_j' is called the relative degree of grey incidence and denoted r_{ij}.

The relative degree of incidence is a quantitative representation of the rates of change of the sequence X_i and X_j relative to their initial values. The closer the rates of change of X_i and X_j are, the larger r_{ij} is and vice versa. ■

Proposition 2.3

Assume that X_i and X_j are two sequences of the same length with nonzero initial values. Their relative degree r_{ij} and absolute degree ε_{ij} of incidence do not require connection. When ε_{ij} is relatively large, r_{ij} may be very small; when ε_{ij} is very small, r_{ij} may also be very large. ■

Proposition 2.4

The relative degree r_{ij} of grey incidence satisfies the following properties:

(i) $0 < r_{ij} \leq 1$;
(ii) The value of r_{ij} relates only to the rates of change of X_i and X_j with respect to their individual initial values and is not connected with the magnitudes

of other entries. In other words, scalar multiplication does not change the relative degree of grey incidence;

(iii) The closer the individual rates of change of X_i and X_j with respect to their initial points, the greater r_{ij} is;

(iv) When $X_j = aX_i$, or when the images of zero initial points of the initial images of X_i and X_j satisfy $X_j^{\prime 0}$ waves around $X_i^{\prime 0}$, and the area of the parts with $X_j^{\prime 0}$ above $X_i^{\prime 0}$ equals that of the parts with $X_j^{\prime 0}$ underneath $X_i^{\prime 0}$, $\rho_{ii} = \rho_{jj} = 1$;

(v) $r_{ii} = r_{jj} = 1$;

(vi) $r_{ij} = r_{ji}$.

Example 2.2

Calculate the relative degree of incidence of Example 2.1:

(1) Obtain the initial image of X_0 and X_1,

$$X_0' = (1, 0.9, 1.5, 1.4, 1.4, 1.5, 1.6)$$

$$X_1' = (1, 1.26, 1.52, 1.67, 1.83, 1.98, 2.13)$$

(2) Find the images of zero starting point, of X_0' and X_1',

$$x_0'^0 = \left(x_0'^0(1), x_0'^0(2), x_0'^0(3), x_0'^0(4), x_0'^0(5), x_0'^0(6), x_0'^0(7) \right)$$

$$= (0, -0.1, 0.5, 0.4, 0.4, 0.5, 0.6)$$

$$x_1'^0 = \left(x_1'^0(1), x_1'^0(2), x_1'^0(3), x_1'^0(4), x_1'^0(5), x_1'^0(6), x_1'^0(7) \right)$$

$$= (0, 0.26, 0.52, 0.67, 0.83, 0.98, 1.13)$$

(3) Calculate $|s_0'|, |s_1'|, |s_1' - s_0'|$

$$|s_0'| = \left| \sum_{k=2}^{6} x_0'^0(k) + \frac{1}{2} x_0'^0(7) \right| = 2$$

$$|s_1'| = \left| \sum_{k=2}^{6} x_1'^0(k) + \frac{1}{2} x_1'^0(7) \right| = 3.828$$

$$|s_1' - s_0'| = \left| \sum_{k=2}^{6} \left(x_1'^0(k) - x_0'^0(k) \right) + \frac{1}{2} \left(x_1'^0(7) - x_0'^0(7) \right) \right| = 1.925$$

(4) Calculate the relative degree of grey incidence

$$r_{01} = \frac{1+|s_0'|+|s_1'|}{1+|s_0'|+|s_1'|+|s_1'-s_0'|} = \frac{6.825}{8.75} \approx 0.78$$

2.2.3 Synthetic Grey Incidence Model

Definition 2.6

Let X_i and X_j be sequences of the same length with nonzero initial entries, ε_{ij} and r_{ij} be, respectively, the absolute and relative degrees of incidence between X_i and X_j, and $\theta \in [0,1]$, then

$$\rho_{ij} = \theta\varepsilon_{ij} + (1-\theta)r_{ij} \qquad (2.9)$$

is the synthetic degree of grey incidence between X_i and X_j (the short form is *synthetic degree of incidence*).

The synthetic degree of grey incidence represents the similarity of X_i and X_j and the rates of change of the sequence of X_i and X_j relative to their initial values. It is an index that describes the nearness relationships of sequences. Generally, $\theta = 0.5$; if the study is more concerned about the relationship between the relevant absolute quantities, θ may have a greater value. Conversely, if the focus is on comparing rates of changes, then θ takes a smaller value.

Proposition 2.5

The synthetic degree ρ_{ij} of incidence satisfies the following properties:

  (i) $0 < \rho_{0i} \le 1$;
 (ii) The value of ρ_{ij} is related to the individual observed values of the sequence X_i and X_j as well as the rates of changes of these values with respect to their initial values;
(iii) When $\theta = 1$, $\rho_{ij} = \varepsilon_{ij}$, when $\theta = 0$, $\rho_{ij} = r_{ij}$;
 (iv) $\rho_{00} = \rho_{ii} = 1$;
  (v) $\rho_{ij} = \rho_{ji}$

Example 2.3

Calculate the synthetic degree of incidence of X_0 and X_1 from Example 2.1.

Solution

From Examples 2.1 and 2.2, take $\varepsilon_{01} = 0.5581$, $r_{01} = 0.78$, choose $\theta = 0.5$, and then

$$\rho_{01} = \theta\varepsilon_{01} + (1-\theta)r_{01} = 0.5 \times 0.5581 + 0.5 \times 0.78 \approx 0.669$$

Table 2.1   Grey Comprehensive Correlation

Value θ	0.2	0.3	0.4	0.6	0.8
Synthetic degree of incidence	0.73562	0.71343	0.69124	0.64686	0.60248

Similarly, if we choose $\theta = 0.2, 0.3, 0.4, 0.6, 0.8$, we can calculate the synthetic degree of grey incidence, as Table 2.1 shows.

2.3 Grey Incidence Models Based on Similarity and Nearness

Analysis of the relationships of curves can be studied from the perspectives of similarity and nearness. Grey incidence models based on these factors are covered in this section.

Definition 2.7

Let X_i and X_j be sequences of the same length, and s_i-s_j the same as defined in Equation (2.2); then

$$\varepsilon_{ij} = \frac{1}{1+|s_i - s_j|} \tag{2.10}$$

Represents the similitude degree of incidence between X_i and X_j or *similitude degree of incidence*. Similitude is employed to measure the geometric similarities of the shapes of sequence X_i and X_j. The more similarity the geometric shapes of X_i and X_j, the greater value ε_{ij} takes, and vice versa.   ■

Definition 2.8

Let X_i and X_j be sequences of the same length, and S_i-S_j the same as defined in Equation (2.3); then

$$\delta_{ij} = \frac{1}{1+|S_i - S_j|} \tag{2.11}$$

represents the nearness degree of grey incidence of X_i and X_j or the close degree of grey incidence. The close degree of incidence is used to measure the spatial nearness of sequence X_i and X_j. The closer X_i and X_j are, the larger the value δ_{ij} takes and vice versa. This concept is meaningful only when the sequences X_i and X_j possess similar meanings and identical units. It has no other practical significance.

A comparison of Equations (2.4) and (2.10) shows that the similitude degree of incidence and the absolute degree of incidence follow the same principle.   ■

Proposition 2.6

The nearness degree δ_{ij} of incidence satisfies the following properties:

(i) $0 < \delta_{ij} \leq 1$;
(ii) The value of δ_{ij} is related to both the geometric shape of the sequence X_i and X_j and their relative spatial positions;
(iii) The closer X_i and X_j are, the greater value δ_{ij} takes, and vice versa;
(iv) When X_i is superpositioned with X_j or vibrates around X_j and the area of the parts where X_i is located above X_j equals the area of the parts where X_i is located beneath X_j, $\delta_{ij} = 1$;
(v) $\delta_{ii} = \delta_{jj} = 1$;
(vi) $\delta_{ij} = \delta_{ji}$.

When computing the similitude degrees or close degrees of incidence based on Equations (2.10) or (2.11) and the absolute values of sequence data are relative large, the values of both $|s_i - s_j|$ and $|S_i - S_j|$ may also be large and the resultant similitude and close degrees of incidence may be relatively small. This scenario does not substantially impact the analysis of order relationships. If an analysis demands relatively large numerical values of the degrees of incidence, one can consider replacing the 1s in the numerators and denominator in Equations (2.10) or (2.11) by a constant related to $|s_i - s_j|$ and $|S_i - S_j|$ use the grey absolute degree of incidence or other appropriate model.   ■

2.4 Grey Evaluation Using Triangular Whitenization Functions

The grey cluster evaluation model based on end-point triangular whitenization function is applicable when the boundary of each grey class is clear and the most possible point of each grey class is unknown. The grey cluster evaluation based on center-point triangular whitenization function is applicable when the most possible point of each grey class is known but the boundary of each grey class is unclear. The triangular whitenization weight function of moderate measure is introduced first.

2.4.1 Triangular Whitenization Function of Moderate Measure

Assume that n objects are to be clustered into s different grey classes according to m evaluation criteria. We classify the i-th object into the k-th grey class according to the observed value x_{ij} ($i = 1,2,...,n$; $j = 1,2,...,m$) of the i-th ($i = 1,2,...,n$) object judged against the k-th ($k \in \{1,2,...,m\}$) criterion. Classifying n objects into s grey classes using the j-th criterion is known grey clustering. The whitenization weight functions of the j-th criterion k-th subclass are denoted as $f_j^k(\cdot)$.

Figure 2.1 shows a moderate measure of triangular whitening weight function. By comparing a typical trapezoidal whitening weight function, we see that the turning points 2 and 3 of the function, usually denoted $f_j^k\left(x_j^k(1), x_j^k(2), -, x_j^k(4)\right)$, overlap.

Assuming x is a j-th criterion, it is not difficult to calculate the membership degree of grey class $k(k = 1,2,...,s)$ from (2.12).

$$
f_j^k(x) =
\begin{cases}
0 & x \notin \left[x_j^k(1), x_j^k(4) \right] \\[2mm]
\dfrac{x - x_j^k(1)}{x_j^k(2) - x_j^k(1)} & x \in \left[x_j^k(1), x_j^k(2) \right] \\[2mm]
\dfrac{x_j^k(4) - x}{x_j^k(4) - x_j^k(2)} & x \in \left[x_j^k(2), x_j^k(4) \right]
\end{cases}
\tag{2.12}
$$

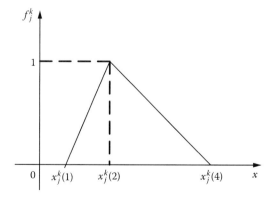

Figure 2.1   Triangular whitening weight function.

2.4.2 Evaluation Model Using End-Point Triangular Whitenization Functions

The computational steps of the grey cluster evaluation model using end-point triangular whitenization weight functions are:

Step 1: Based on a predetermined number s of grey classes, divide the individual ranges of the criteria into s grey classes. For example, let $[a_1, a_{s+1}]$ be the range of the values of criterion j. Now, divide $[a_1, a_{s+1}]$ into s grey classes as

$$[a_1, a_2], \cdots, [a_{k-1}, a_k], \cdots, [a_{s-1}, s_s], [a_s, a_{s+1}]$$

where a_k ($k = 1, 2, \cdots, s, s + 1$) can be determined based on specific requirements or relevant qualitative analysis.

Step 2: Calculate the geometric midpoints between the cells, $\lambda_k = (a_k + a_{k+1})/2$, $k = 1, 2, \cdots, s$.

Step 3: Let the whitenization weight function value for λ_k to be in the k-th grey class be 1. When $(\lambda_k, 1)$ is connected to the midpoint λ_{k-1} of the $(k - 1)$-th grey class and the midpoint λ_{k+1} of the $(k + 1)$-th grey class, one obtains a triangular whitenization weight function $f_j^k(\cdot)$ in terms of criterion j about the k-th grey class, $j = 1, 2, \cdots, m$; $k = 1, 2, \cdots, s$. For $f_j^1(\cdot)$ and $f_j^s(\cdot)$, the range of criterion j can be extended to the left and right to a_0 and a_{s+2}, respectively (see Figure 2.2). For any observed value x of criterion j, its degree $f_j^k(x)$ of membership in the k-th grey class, $j = 1, 2, \cdots, m$; $k = 1, 2, \cdots, s$, can be computed:

$$f_j^k(x) = \begin{cases} 0, & x \notin [\lambda_{k-1}, \lambda_{k+1}] \\ \dfrac{x - \lambda_{k-1}}{\lambda_k - \lambda_{k-1}}, & x \in [\lambda_{k-1}, \lambda_k] \\ \dfrac{\lambda_{k+1} - x}{\lambda_{k+1} - \lambda_k}, & x \in [\lambda_k, \lambda_{k+1}] \end{cases} \tag{2.13}$$

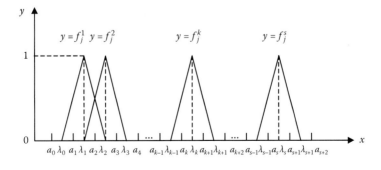

Figure 2.2   Construction of triangular whitenization weight function.

Step 4: Obtain the weight η_j, $j = 1,2,\cdots,m$ for each index in comprehensive clustering.
Step 5: Compute the comprehensive clustering coefficient σ_i^k for object $i(i = 1,2,\cdots,n)$ in terms of the k-th grey class, $k = 1,2,\cdots,s$

$$\sigma_i^k = \sum_{j=1}^{m} f_j^k(x_{ij}) \cdot \eta_j \qquad (2.14)$$

where $f_j^k(x_{ij})$ is the whitenization weight function of the k-th subclass of the j-th criterion, and η_j the weight of criterion j in the comprehensive clustering.

Step 6: From $\max_{1 \le k \le s}\{\sigma_i^k\} = \sigma_i^{k^*}$, it follows that object i belongs to the k^* grey class. When several objects belong to the same k^* grey class, one can further determine the order of preference among them based on the magnitudes of their individual cluster coefficients.

2.4.3 Evaluation Model Using Center-Point Triangular Whitenization Functions

The point of maximum greyness is called the center of the class. The specific steps to use the grey evaluation model based on the center-point triangular whitenization weight functions are:

Step 1: Based on the number s of grey classes required by the evaluation task, determine the center points $\lambda_1,\lambda_2,\cdots,\lambda_s$ of grey classes $1,2,\cdots,s$, respectively. Then divide the individual ranges of the criteria into s grey classes. For example, let $[\lambda_1,\lambda_{s+1}]$ be the range of the values of criterion j, then divide $[\lambda_1,\lambda_{s+1}]$ into s grey classes as follows:

$$[\lambda_1,\lambda_2],\cdots,[\lambda_{k-1},\lambda_k],\cdots,[\lambda_{s-1},\lambda_s],[\lambda_s,\lambda_{s+1}]$$

Step 2: Connect$(\lambda_k,1)$ with the $(k-1)$-th center point $(\lambda_{k-1},0)$of a small range and the $(k+1)$-th center point $(\lambda_{k+1},0)$of a subinterval to find the triangular whitenization weight function $f_j^k(\cdot)$on the j-th criterion to the k grey class, $j = 1,2,\cdots,m$; $k = 1,2,\cdots,s$. For $f_j^1(\cdot)$ and $f_j^s(\cdot)$, the number of fields of the j-th criterion can be extended to λ_0, λ_{s+1}, respectively, to the left and right; see Figure 2.3. For an observed value x of criterion j, we can employ

$$f_j^k(x) = \begin{cases} 0, & x \notin [\lambda_{k-1},\lambda_{k+1}] \\ \dfrac{x - \lambda_{k-1}}{\lambda_k - \lambda_{k-1}}, & x \in (\lambda_{k-1},\lambda_k] \\ \dfrac{\lambda_{k+1} - x}{\lambda_{k+1} - \lambda_k}, & x \in (\lambda_k,\lambda_{k+1}) \end{cases} \qquad (2.15)$$

to calculate its degree of membership $f_j^k(x)$ in grey class $k(k = 1,2,\cdots,s)$.

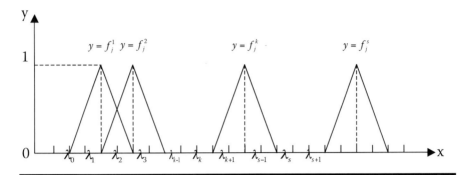

Figure 2.3 Center triangular whitening weight function.

Step 3: Obtain the weight $\eta_j, = j = 1,2,\cdots,m$ for each index in comprehensive clustering.

Step 4: Compute the comprehensive clustering coefficient σ_i^k for object $i(i = 1,2,\cdots,n)$ with respect to grey class $k(k = 1,2,\cdots,s)$:

$$\sigma_j^k = \sum_{j=1}^{m} f_j^k(x_{ij}) \cdot \eta_j \tag{2.16}$$

where $f_j^k(x_{ij})$ is the whitenization weight function of the k-th subclass of the j-th criterion, and η_j the weight of criterion j in the comprehensive clustering.

Step 5: From $\max\limits_{1 \leq k \leq s}\{\sigma_i^k\} = \sigma_i^{k^*}$, it follows that object i belongs to the k^* grey class. When the grey class k^* contains several objects, they can be ordered according to the magnitudes of their comprehensive clustering coefficients.

2.5 Multiattribute Grey Target Decision Model

Grey target decisions involve the application of the principle of nonuniqueness in decision making and is mainly suitable for objectives with satisfactory solutions. Grey target decision models are widely used by the petroleum industry, the military, and various engineering fields. This section constructs four new functions of uniform effect measures to establish a new multiattribute grey target decision model. The cases of hitting or not hitting the bull's eye of the objective effect value are fully considered. The model greatly improves the ability to distinguish synthetic effect measures.

2.5.1 Basic Concepts

Definition 2.9

The totality (or set) of events within a range of research is denoted $A = \{a_1, a_2, \cdots, a_n\}$

where a_i ($i = 1,2,3, \cdots,n$) is the i-th event. The corresponding totality (set) of all the possible countermeasures is denoted $B = \{b_1,b_2,\cdots,b_m\}$ where $b_j(j = 1,2,\ldots m)$ is the j-th countermeasure.   ■

Definition 2.10

Assume that $A = \{a_1,a_2,\ldots a_n\}$ is a set of research events and $B = \{b_1,b_2,\cdots,b_m\}$ is the countermeasure set. Then the Cartesian product $A \times B = \{(a_i, \ b_j)|a_i\in A, \ b_j\in B\}$ is called the situation set, denoted $S = A \times B$. For any $a_i\in A$, $b_j\in B$, the pair $(a_i, \ b_j)$ is called a situation, denoted $s_{ij} = (a_i, \ b_j)$.   ■

Definition 2.11

Assume that $A = \{a_1,a_2,\ldots a_n\}$ is the set of events, and $B = \{b_1,b_2,\cdots,b_m\}$ is the countermeasure set, $S = \{s_{ij} = (a_i, \ b_j)|a_i\in A,b_j\in B\}$ is the situation set, $u_{ij}^{(k)}$ the effect value of situation s_{ij} with objective k, and R the set of all real numbers. Then $u_{ij}^{(k)} : s \mapsto R$.

$$s_{ij} \mapsto u_{ij}^{(k)}$$

is called the effect mapping of S with objective k. The target effect value $u_{ij}^{(k)}$ is embodied for objective achievement or a concrete embodiment of the degree of effective achievement or deviation.   ■

Definition 2.12

Assume that $d_1^{(1)}, d_2^{(1)}$ are the upper and the lower threshold values, respectively, for situation effects with objective k. $S^1 = \left\{r \middle| d_1^{(k)} \leq r \leq d_2^{(k)}\right\}$ is the grey target of one-dimensional decision making with objective k, $u_{ij}^{(k)} \in \left[d_1^{(k)}, d_2^{(k)}\right]$ a satisfactory effect with objective k, the corresponding s_{ij} the desirable situation with objective k, and b_j the desirable countermeasure with respect to event a_i with objective k.   ■

Definition 2.13

Assume that $d_1^{(1)}, d_2^{(1)}$ are the threshold values of situation effects for objective 1, and $d_1^{(2)}, d_2^{(2)}$, the threshold values of situation effects for objective 2. Then

$$S^2 = \left\{(r^{(1)},r^{(2)}) \middle| d_1^{(1)} \leq r^{(1)} \leq d_2^{(1)}, d_1^{(2)} \leq r^{(2)} \leq d_2^{(2)}\right\}$$

is a grey target of two-dimensional decision making. If the effect vector of the situation s_{ij} satisfies $u_{ij} = \left\{ u_{ij}^{(1)}, u_{ij}^{(2)} \right\} \in S^2$, then s_{ij} is said to be a desirable situation with objectives 1 and 2 and b_j a desirable countermeasure of the event a_i with objectives 1 and 2. ■

Definition 2.14

Assume that $d_1^{(1)}, d_2^{(1)}, d_1^{(2)}, d_2^{(2)}; \cdots; d_1^{(s)}, d_2^{(s)}$ are the threshold values of situation effects with objectives $1,2,\ldots,s$, respectively. The following region of the s-dimensional Euclidean space

$$ S^s = \left\{ (r^{(1)}, r^{(2)}, \cdots, r^{(s)}) \middle| d_1^{(1)} \le r^{(1)} \le d_2^{(1)}, d_1^{(2)} \le r^{(2)} \le d_2^{(2)}, \cdots d_1^{(s)} \le r^{(s)} \le d_2^{(s)} \right\} $$

is the grey target of s-dimensional decision making. If the effect vector of the situation s_{ij} satisfies $u_{ij} = \left(u_{ij}^{(1)}, u_{ij}^{(2)}, \cdots, u_{ij}^{(s)} \right) \in S^s$ where $u_{ij}^{(k)} (k = 1, 2, \cdots, s)$ is the effect value of the situation s_{ij} with objective k, then s_{ij} is a desirable situation with objective $1,2,\ldots,s$, b_j a desirable countermeasure of the event a_i with objective $1,2,\ldots,s$.

The grey target decision model is essentially intended to determine relative optimum results. Absolute best results are impossible to achieve so organizations try to find satisfactory results in which the target effect vector s_{ij} requires $u_{ij} = \left(u_{ij}^{(1)}, u_{ij}^{(2)}, \cdots, u_{ij}^{(s)} \right) \in S^s$ ("hitting the target"). ■

2.5.2 Construction of Matrix of Uniform Effect Measures

The significance of target effect values, dimensions, and natures may vary. To obtain a combined measure of comparable situation effects, we first turn the effect value $u_{ij}^{(k)}$ to the same effect measure.

Definition 2.15

Assume that $A = \{a_1, a_2, \ldots a_n\}$ is the set of events, $B = \{b_1, b_2, \ldots b_m\}$ is the countermeasure set, $S = \{s_{ij} = (a_i, b_j) | a_i \in A, b_j \in B\}$ is the situation set,

$$ U^{(k)} = \left(u_{ij}^{(k)} \right) = \begin{bmatrix} u_{11}^{(k)} & u_{12}^{(k)} & \cdots & u_{1m}^{(k)} \\ u_{21}^{(k)} & u_{22}^{(k)} & \cdots & u_{2m}^{(k)} \\ \cdots & \cdots & \cdots & \cdots \\ u_{n1}^{(k)} & u_{n2}^{(k)} & \cdots & u_{nm}^{(k)} \end{bmatrix} $$

is the effect sample matrix of situation set S with objective k ($k = 1, 2, \cdots, s$). Assume:

(1) k is a benefit type objective, that is, a larger objective effect sample value is better; assume $u_{ij}^{(k)} \in \left[u_{i_0 j_0}^{(k)}, \max_i \max_j \left\{ u_{ij}^{(k)} \right\} \right]$ is the target with objective k, that is, $u_{i_0 j_0}^{(k)}$ is the threshold of objective sample, then

$$r_{ij}^{(k)} = \frac{u_{ij}^{(k)} - u_{i_0 j_0}^{(k)}}{\max_i \max_j \left\{ u_{ij}^{(k)} \right\} - u_{i_0 j_0}^{(k)}} \tag{2.17}$$

is the effect measure.

(2) k is a cost type objective, that is, a lower objective effect sample value is better; assume $u_{ij}^{(k)} \in \left[\min_i \min_j \left\{ u_{ij}^{(k)} \right\}, u_{i_0 j_0}^{(k)} \right]$ is the target with objective k, that is, $u_{i_0 j_0}^{(k)}$ is the threshold value of objective sample, then

$$r_{ij}^{(k)} = \frac{u_{i_0 j_0}^{(k)} - u_{ij}^{(k)}}{u_{i_0 j_0}^{(k)} - \min_i \min_j \left\{ u_{ij}^{(k)} \right\}} \tag{2.18}$$

is the effect measure.

(3) k is a moderate type objective, that is, an objective effect sample value closer to the moderate value A is better. Assume $u_{ij}^{(k)} \in [A - u_{i_0 j_0}^{(k)}, A + u_{i_0 j_0}^{(k)}]$ is the target with objective k, that is $A - u_{i_0 j_0}^{(k)}$, $A + u_{i_0 j_0}^{(k)}$, respectively, stands for the threshold values of lower and upper effects, then

(a) If $u_{ij}^{(k)} \in [A - u_{i_0 j_0}^{(k)}, A]$,

$$r_{ij}^{(k)} = \frac{u_{ij}^{(k)} - A + u_{i_0 j_0}^{(k)}}{u_{i_0 j_0}^{(k)}} \tag{2.19}$$

is called the lower effect measure.

(b) If $u_{ij}^{(k)} \in [A, A + u_{i_0 j_0}^{(k)}]$,

$$r_{ij}^{(k)} = \frac{A + u_{i_0 j_0}^{(k)} - u_{ij}^{(k)}}{u_{i_0 j_0}^{(k)}} \tag{2.20}$$

is called the upper effect measure. ■

The effect measures of benefit type objectives reflect the how close effect sample values are to maximum sample values and how far they are from the threshold effect values of the objectives. Similarly, the effect measures of cost type objectives show how close the effect sample values are to the minimum effect sample values and how far the effect sample values are from the threshold effect values of the objectives. The lower effect measures of moderate value type objectives indicate how close the effect sample values smaller than the moderate value A are to the moderate value A and how far they are from the lower threshold effect values of the objectives. The upper effect measures of moderate value type objectives indicate how close the effect samples larger than the moderate value A are to the moderate value A and how far they are from the upper threshold effect values of the objectives. Out-of-target (undershoot) cases are classified four ways:

(1) Benefit type objective value is less than critical value $u_{i_0 j_0}^{(k)}$, that is, $u_{ij}^{(k)} < u_{i_0 j_0}^{(k)}$;

(2) Cost type objective value is larger than critical value $u_{i_0 j_0}^{(k)}$, that is, $u_{ij}^{(k)} > u_{i_0 j_0}^{(k)}$;

(3) Moderate type objective value is less than lower effect critical value $A - u_{i_0 j_0}^{(k)}$, that is, $u_{ij}^{(k)} < A - u_{i_0 j_0}^{(k)}$;

(4) Moderate type objective value is larger than upper effect critical value $A + u_{i_0 j_0}^{(k)}$, that is, $u_{ij}^{(k)} > A + u_{i_0 j_0}^{(k)}$.

To determine the effect measure of each type of objective to satisfy the condition of normality, that is, $r_{ij}^{(k)} \in [-1,1]$, without loss of generality:

For the benefit type objective, assume $u_{ij}^{(k)} \geq - \max_i \max_j \left\{ u_{ij}^{(k)} \right\}, 2u_{i_0 j_0}^{(k)}$;

For the cost type objective, assume $u_{ij}^{(k)} \geq - \min_i \min_j \left\{ u_{ij}^{(k)} \right\} + 2u_{i_0 j_0}^{(k)}$;

When the moderate type objective value is less than the lower effect critical value $A - u_{i_0 j_0}^{(k)}$, assume $u_{ij}^{(k)} \geq A - 2u_{i_0 j_0}^{(k)}$;

When the moderate type objective value is larger than the upper effect critical value $A + u_{i_0 j_0}^{(k)}$, assume $u_{ij}^{(k)} \leq A + 2u_{i_0 j_0}^{(k)}$;

This leads to Proposition 2.7.

Proposition 2.7

The three effect measures $r_{ij}^{(k)} (i = 1, 2, \cdots, n; j = 1, 2 \cdots, m; k = 1, 2, \cdots, s)$ as given by Definition 2.15 satisfy the following concepts:

(i) $r_{ij}^{(k)}$ has no dimension;

(ii) The more ideal the effect is, the greater $r_{ij}^{(k)}$ is;

(iii) $r_{ij}^{(k)} \in [-1,1]$.

If the effect measure "hits the bull's eye," $r_{ij}^{(k)} \in [0,1]$; otherwise, $r_{ij}^{(k)} \in [-1,0]$. ■

Definition 2.16

The effect measures for the benefit type and cost type objectives and the lower and upper effect measures for the moderate type $r_{ij}^{(k)}$ $(i = 1, 2, \cdots, n; j = 1, 2 \cdots, m; k = 1, 2, \cdots, s)$ are all called uniform effect measures; they reveal the achievement of or deviation from each objective. For a benefit type objective, expect to achieve a goal of "the bigger the better" or "the fewer the better." For a cost type objective, the goal is "the smaller the better" or "the more the better." The goal for a moderate type objective is "neither too large nor too small" or "neither too much nor too little." ■

2.5.3 Construction of Matrix of Synthetic Effect Measures

Definition 2.17

Assume $\eta_k (k = 1, 2, \cdots, s)$ is the decision weight of objective k, satisfying $\sum_{k=1}^{s} \eta_k = 1$. Then

$$R^{(k)} = \left(r_{ij}^{(k)} \right) = \begin{bmatrix} r_{11}^{(k)} & r_{12}^{(k)} & \cdots & r_{1m}^{(k)} \\ r_{21}^{(k)} & r_{22}^{(k)} & \cdots & r_{2m}^{(k)} \\ \cdots & \cdots & \cdots & \cdots \\ r_{n1}^{(k)} & r_{n2}^{(k)} & \cdots & r_{nm}^{(k)} \end{bmatrix}$$

is called the matrix of uniform effect measure of the situation set S with objective k. Suppose that $s_{ij} \in S$, then

$$r_{ij} \sum_{k=1}^{s} \eta_k \cdot r_{ij}^{(k)} \tag{2.21}$$

is called the synthetic effect measure of the situation s_{ij} and

$$R = (r_{ij}) = \begin{bmatrix} r_{11} & r_{12} & \cdots & r_{1m} \\ r_{21} & r_{22} & \cdots & r_{2m} \\ \cdots & \cdots & \cdots & \cdots \\ r_{n1} & r_{n2} & \cdots & r_{nm} \end{bmatrix}$$

is called the matrix of synthetic effect measures. ■

Proposition 2.8

The synthetic effect measure $r_{ij}(i = 1,2,\cdots,n; j = 1,2,\cdots,m)$ as given in Definition 2.17 satisfies the following conditions:

(i) r_{ij} has no dimension;
(ii) The more ideal the effect is, the greater r_{ij} is;
(iii) $r_{ij} \in [-1,1]$.

If the synthetic effect measure hits the bull's eye, $r_{ij} \in [0,1]$; otherwise, $r_{ij} \in [-1,0]$. When the synthetic effect measure hits the bull's eye, by comparing $r_{ij}(i = 1,2,\cdots, n; j = 1,2,\cdots,m)$ we can judge the merits of event $a_i(i = 1,2,\cdots,m)$, countermeasure $b_j(j = 1,2,\cdots,n)$ and the $s_{ij}(i = 1,2,\cdots,n; j = 1,2,\cdots,m)$ situation. ■

Definition 2.18

(1) If $\max\limits_{1\leq j\leq m}\{r_{ij}\} = r_{ij_0}$, then b_{j_0} is called the optimum countermeasure of event a_i;

(2) If $\max\limits_{1\leq j\leq m}\{r_{ij}\} = r_{i_0 j}$, then a_{i_0} is called the optimum event corresponding to the countermeasure b_j;

(3) If $\max\limits_{1\leq i\leq m}\max\limits_{1\leq j\leq m}\{r_{ij}\} = r_{i_0 j_0}$, then $s_{i_0 j_0}$ is called the optimum situation. ■

2.5.4 Steps of Multiattribute Grey Target Assessment Algorithm

The steps of multiattribute grey target assessment algorithm are:

Step 1: Construct the situation set $S = \{s_{ij} = (a_i,b_j)|a_i \in A, b_j \in B\}$ according to the events set $A = \{a_1,a_2,\cdots,a_n\}$ and the countermeasures set $B = \{b_1,b_2,\cdots,b_m\}$.
Step 2: Fix the decision objective $k = 1,2,\cdots,s$.
Step 3: Determine the weight $\eta_1,\eta_2,\cdots,\eta_s$ for each objective.
Step 4: Solve the matrix of effect sample corresponding to objective $k = 1,2,\cdots,s$.

$$U^{(k)} = \left(u_{ij}^{(k)}\right) = \begin{bmatrix} u_{11}^{(k)} & u_{12}^{(k)} & \cdots & u_{1m}^{(k)} \\ u_{21}^{(k)} & u_{22}^{(k)} & \cdots & u_{2m}^{(k)} \\ \cdots & \cdots & \cdots & \cdots \\ u_{n1}^{(k)} & u_{n2}^{(k)} & \cdots & u_{nm}^{(k)} \end{bmatrix}$$

Step 5: Set the threshold value of objective effect.
Step 6: Solve the matrix of uniform effect measure with objective k.

$$R^{(k)} = \left(r_{ij}^{(k)} \right) = \begin{bmatrix} r_{11}^{(k)} & r_{12}^{(k)} & \cdots & r_{1m}^{(k)} \\ r_{21}^{(k)} & r_{22}^{(k)} & \cdots & r_{2m}^{(k)} \\ \cdots & \cdots & \cdots & \cdots \\ r_{n1}^{(k)} & r_{n2}^{(k)} & \cdots & r_{nm}^{(k)} \end{bmatrix}$$

Step 7: Calculate the matrix of synthetic effect measures

$$R = \left(r_{ij} \right) = \begin{bmatrix} r_{11} & r_{12} & \cdots & r_{1m} \\ r_{21} & r_{22} & \cdots & r_{2m} \\ \cdots & \cdots & \cdots & \cdots \\ r_{n1} & r_{n2} & \cdots & r_{nm} \end{bmatrix}$$

by $r_{ij} = \sum_{k=1}^{s} \eta_k \cdot r_{ij}^{(k)}$.

Step 8: According to Definition 2.18, define the optimum countermeasure b_{j_0}
or the optimum situation $s_{i_0 j_0}$.

Chapter 3

Postevaluation of Road–Bridge Construction: Case Study of Lianxu Highway in China

3.1 Introduction

3.1.1 Postevaluation

Postevaluation involves the performance and efficiency analysis of the construction of a project and benefit assessment. Postevaluation of projects started in the 1930s. In recent decades, postevaluation became common practice for national governments and financial organizations such as the World Bank and the Asian Development Bank that use the technique to improve investment efficiency. Postevaluation in China started in the 1980s and focused on new systems and evaluation approaches. However, research is still in the preliminary stages in relation to road and bridge construction projects.

The principles and traits of project postevaluation are different from those used for feasibility, preevaluation, midevaluation, and completion assessments. These types of assessments are related to but cannot replace postevaluation.

3.1.1.1 Comparison of Feasibility Evaluation and Preevaluation

Feasibility and preevaluation studies analyze technical advantages, economic efficiency, and project feasibility after careful investigation and forecasting are completed. They pave the way for project decision making. Postevaluation has the following features compared to feasibility evaluation and preevaluation:

1. Reality: Postevaluation focuses on analysis of status quo. It is a reforecast based on real performance after a project has been in operation for several years. Feasibility analysis and preevaluation utilize forecasts based on historical data or experience.
2. Comprehensiveness: All aspects of a project including investment issues, operation, and economic efficiency should be evaluated.
3. Exploration: It is crucial to use postevaluation to analyze performance of a project to find shortcomings and direct future development. Very creative evaluators are required to assess project efficiency and propose improvements.
4. Feedback: Feasibility evaluation and preevaluation are conducted to facilitate decision making. Postevaluation aims to provide feedback to concerned parties and verify investment decisions.
5. Cooperation. Evaluators and investors should cooperate during feasibility evaluation and preevaluation and generate assessment reports. In contrast, postevaluation requires broad collaboration by technical personnel, project managers, enterprise managers, and investment specialists.

3.1.1.2 Comparison with Midevaluation

Midevaluation involves a comparison of forecasts and plans to current performance to achieve project improvements. Midevaluation and postevaluation differ in several important areas:

Project management: Midevaluation is ongoing during construction; postevaluation is conducted after completion.

Purpose and function: Midevaluation examines discrepancies between recent performance and forecasts and provides feedback to management to improve construction performance. Postevaluation reflects discrepancies over the life of a project.

Implementation: Midevaluation is usually performed by departments or management responsible for project management; postevaluation is usually carried by an independent organization.

Evaluation content: Midevaluation focuses on construction performance by analyzing gaps between progress and forecast schedules, reasons for cost deficits, and contractor performance. Postevaluation has a wider focus and evaluates every aspect of a project already in operation.

3.1.1.3 Comparison of Acceptance and Audit

Conditions for acceptance of a project after completion are defined in design documents and specify expected results. Acceptance is an important aspect of postevaluation. Assessment of investments in fixed assets at final acceptance is one of the first tasks of postevaluation.

A project audit focuses on compliance with regulations, financial wastes and losses, and investment activities. After an audit, the financial data from a project is more reliable. The disclosure of wastes and losses later serves as important data for postevaluation. An extended project audit also evaluates decision making, design, procurement, final acceptance, and benefits.

Postevaluation of a road–bridge project requires extensive observation along with accurate operation data obtained from completion assessments, audits, and midevaluations.

3.1.2 Lianxu Highway Project

3.1.2.1 Overview of Project

3.1.2.1.1 Brief Description of Project

The Lianhuo Highway across central China links the East, Central Plain, and Northwest parts of the country. It is a corridor to the Pacific and an important part of the national highway system (Figure 3.1). The Lianxu Highway is the eastern segment of the Lianhuo Highway, starting in Xuzhou, Lianyungang and terminating in Laoshankou, Xuzhou at the boundary of the Jiangsu and Anhui Provinces. Its overall length is 236 km and with national highways 204, 205, 206, and 104, constitutes the local highway network.

The National Department of Transportation conducted prefeasibility studies and the report was completed in December 1993. In May 1994, the feasibility

Figure 3.1   Lianxu highway.

studies were completed and the report passed review by the Jiangsu Transportation Bureau in August 1994. The project underwent assessment of the feasibility report in 1996 and modifications due to changes of the highway structure were made in 1996 and 1997. The report with supplements was finalized in March 1998. The fundamental design was prepared in 1994 and redesign based on the requirements of the new report started in 1996. The project was approved by the National Development and Reform Commission, the Ministry of Transport, and the Ministry of Environmental Protection in 1998.

The design of the first stage of the Lianxu Highway started in March 1996 and was completed in 1997 and modified at the direction of the Jiangsu Highway Construction Headquarters in August 1998. The design of the section from the start point to Dadaoshan was again revised after blueprints of the first phase were submitted. The design of the second phase started in August 1997 and was completed in August 1998. The viaduct section designs were completed in May 1999.

After blueprints for the first stage were submitted, a viaduct design was required and completed in May 1999 after investigation of the section from the start point to Dadaoshan. The construction of the Lianxu Highway was a government countermeasure to aid recovery from the Asian financial crisis of 1997–1998. Construction of the first stage started in March 1997. The experimental section was 14.7 km long.

After the completion of the Lianxu Highway, the acceptance conference for the first stage was held in November 2001. The conference for the second stage in Xuzhou was held in October 2002, and for the second stage in Lianyungang was held in June 2003. On November 29, 2003, the Lianxu Highway project was accepted by the Jiangsu Transport Bureau and Ministry of Transport.

3.1.2.1.2 Investments

The volume of earthwork and stonework totaled 34.0543 million cubic meters. The land requirement was 37031.2 mu. The total expenditure was 6.53 billion yuan.

3.1.2.1.3 Outputs

The two-way, four-lane highway has a length of 236.784 km. Auxiliary facilities along the highway handle communications, tolls, monitoring, and security. The highway has four service areas, two parking areas, and eleven toll facilities. Final acceptance occurred in June 2003. It received an excellent (highest) grade, was named a model road of Jiangsu, and received the Jeme Tien Yow Civil Engineering Award in 2004.

Table 3.1   Main Design Parameters

	Technical Parameter	*Descriptions*
1	Level	Two-way, four-lane highway
2	Traffic capacity	56,100 standard vehicles per day
3	Estimated speed	120 km/h
4	Width of road bed	28 m
	Width of carriage way	2*2*3.75 m
	Width of median	0.75 m+3 m + 0.75 m
	Hard shoulder	3.5 m
	Earth shoulder	0.75 m
	Width of bridge	2* net 12 m
	Surface	Bituminous concrete
	Standard axle load	100 KN
5	Design period	15 years
6	Max. longitudinal gradient	3%
7	Designed bridge load	Cars, 20; trailers, 120
8	Designed flood frequency	Grand bridge 1/300, large, middle, small, culverts 1/100
9	Designed seismic intensity	VII degree

3.1.2.2 Design Parameters

The complete highway length is 236.784 km. The design parameters met the national specifications as shown in Table 3.1.

3.2 Process Evaluation

3.2.1 Preliminary Work and Evaluation

The preconstruction work required four steps: the prefeasibility study, feasibility study, preliminary design, and blueprint design. The preliminary work was discussed in Section 3.1.2. The need for the project was revealed by the feasibility report analyzing the transportation network plan, improvements required, economic development benefits and access to the port of Lianyungang.

3.2.2 Process Design

3.2.2.1 Blueprint Design and General Information

3.2.2.1.1 General Design

The general design considered conditions such as investment level, scale of construction, and design standards. The coordination of the layout and environment factors was considered after careful investigation of existing highway networks, weather conditions, industrial locations, climate, and other factors. The general design also detailed specific design tasks.

3.2.2.1.2 Route Design

The best route was selected by evaluating several options based on technical and economic criteria.

3.2.2.1.3 Pavement Design

The two-way, four lane highway has a roadbed width of 28 m, a curb zone cross slope, carriageway, and hard and soft shoulders. Where the filling height of the embankment is less than 6 m, the slope-by-slope ratio is 1:1.5 where it is higher than 6 m, that of the fundus upper 6 m is 1:1.5, and that below the 6 m is 1:1.75. The ratio of the margin slope of the excavation roadbed is 1:0.75 or 1:1.5, based on rock mass characteristics, height, and the surrounding environment. A combination of natural and geographic conditions required slope protection, drainage ditches and other improvements.

3.2.2.1.4 Bridge and Culvert Design

Based on actual conditions along the proposed highway, review of various types of designs, and a comprehensive analysis of complex geological factors, an appropriate bridge culvert was designed.

3.2.2.1.5 Line Crossover Design

After repeated discussions and reviews during the design process, 11 interchanges, 38 separated interchanges, and 249 channels were designed to meet regional transport needs.

3.2.2.1.6 Traffic Engineering and Facilities

New technologies, techniques and products from China and abroad were used and met the standards for reliable, responsive, accurate, and efficient highway management control.

3.2.2.2 Preparation of Tender Documents

A tender committee jointly established by the Jiangsu Transportation Bureau and the Jiangsu Highway Construction Headquarters was responsible for evaluation of bids. The committee was tasked with preparing tender documents based on Ministry of Transportation standards, and conducting and supervising bidding.

3.2.2.3 Project Implementation and Start of Construction

The state highway construction and development authorities authorized a 5-km test section before issuing a start instruction. From design to implementation, the project required support from road construction workers, local governments and consideration of environmental conditions. The authorities followed inspection and technical standards, quality requirements, environmental protection specifications, and quality standards, water conservation requirements, and safety measures. They also respected local customs along the route. All these factors led to good results.

The first phase of construction was scheduled from 1998 through 2001. A second phase of construction from 2000 to 2003 brought the total to 6 years. The basic construction was completed within the stipulated time, but partial foundation cracking and slip problems extended construction time and increased costs. The project ultimately met the design and specification requirements and achieved a good quality rating. Problems that arose early were resolved.

3.2.2.4 Main Technical Indicators and Evaluation of Changes

During the course of construction, the technical requirements, route length, scale of construction and other parameters changed slightly as shown in Table 3.2.

After the start operations, reasonable changes were made to better accommodate future traffic growth, economic development, and living conditions. Changes improved the network structure, made the road level more reasonable, widened the roadbed, and achieved other improvements as the project advanced.

3.2.3 Implementation and Evaluation of Investment

3.2.3.1 Investment Changes

The total investment projected in the feasibility study exceeded the amount set by the prefeasibility study by 18.02%; the preliminary design budget exceeded the feasibility study budget by 8.50%. The final totals were 17.17% below the preliminary design budget. Figure 3.2 shows investment changes at all stages.

Table 3.2 Highway Construction Change Indicators

Indicators of Change	Prefeasibility	Feasibility		Final Acceptance	
		A	B	A	B
Technical grade	Proposed standards for highway construction	Two-way, four-lane highway with interchanges; capacity: 56,100 pcu/day		Two-way, four-lane highway with interchange; capacity: 56,100 pcu/day	
Total length	238.045 km	Route length 239.485 km		Route length: 236.784 km (3.701 km less than feasibility study length)	
Lengths of stage works		Route length: 108.365 km	B project route length: 131.120 km	Route length of project: 94.12 km; 74.91 km east; 19.21 km west; project feasibility study work to decrease: 14.245 km	Phase II feasibility study work to increase: 11.544 km
Design speed	100 km/h	120 km/h		120 km/h	
Net width of bridge	2'10.75 m	2'12 m		2'12 m	
Subgrade width	24.5 m	28 m		28 m	

Land use	Occupation of land: 9,670 acres	Occupation of land: 10,413 acres	Occupation of land (including land for borrow site): 13,577.07 acres	Occupation of land: 23,454.13 acres
Project scale	Earthwork: 1244 million cubic meters Aasphalt concrete pavement: 231.5385 million square meters	Earthwork: 15.63 million cubic meters Asphalt concrete pavement: 2,849,164 square meters	Earthwork: 12.36 million cubic meters Asphalt concrete pavement: 2,552,200 square meters	Earthwork: 21.696 million cubic meters Asphalt concrete pavement: 3,555,082 square meters
Services	Service (parking) areas: 3 Main toll stations: 2	Services (parking) areas: 2 Main toll stations: 1	Services (parking) areas: 3 Toll stations: 5	Services (parking) areas: 3 Toll stations: 5
Bridges, culverts, cross projects	Bridges: 60 Culverts: 230 Interchanges: 7 Separated interchanges: 18 Channels: 150 Bridges: 6	Bridges: 82 Culverts: 260 Interchanges: 4 Separated interchanges: 12 Channels: 186 Flyovers: 3	Bridges: 69 Culverts: 235 Interchanges: 6 Separated interchanges: 21 Channels: 125	Bridges: 158 Culverts: 232 Interchanges: 5 Separated interchanges: 17 Channels: 132

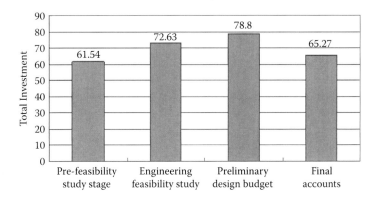

Figure 3.2   Investment changes at various stages.

3.2.3.2 Analysis of Investment Changes

3.2.3.2.1 Engineering Feasibility Study Estimates Exceed Prefeasibility Study Estimates

1. Difference between investment levels cited in prefeasibility study and feasibility study
2. Changes in the scale of construction
3. Changes in duration of construction
4. Increased labor costs
5. Land acquisition price changes
6. Increased interest cost due to changes in funding programs

The prefeasibility report and feasibility report specified a longer completion time during which a number of indicators and methodologies changed. The total investment changed, as did the construction scale, construction time, and other factors. The investment estimated in the feasibility study was greater than the estimate in the prefeasibility study.

3.2.3.2.2 Preliminary Design Budget Estimate Exceeds Feasibility

1. Increased costs of construction and installation.
2. Other construction costs increased 4.2 billion RMB; expense reserves decreased 3.7 billion RMB.
3. The route recommended in the feasibility study was improved in the preliminary design; this increased management conservation, and construction costs.

3.2.3.2.3 Final Budget Less than Preliminary Design Budget

1. Bidding system specification was reasonable. Strict implementation of competitive bidding by suppliers reduced project costs.
2. Economical use of resources and devising new ways to reduce construction costs.
3. Control construction and management expenses by implementing budget management and control systems.
4. Throughout construction, the material prices remained stable to an extent and resulted in savings.

3.2.3.3 Financing Options

Table 3.3 shows financing data. Based on the engineering feasibility study, preliminary design, and the actual financing data, the main changes are described below.

When the original project proposal was approved, Export–Import Bank of Japan loans of $200,000,000 were negotiated. The bank was not interested in completing the loans. Loans from the National Development Bank were obtained. The Ministry of Transportation and Jiangsu Fund for Highway Construction allocated special funds. The total funding was less than the amounts cited in the feasibility study and preliminary design. Bonds were issued to fill the funding gap and ensure smooth progress of the project. City matching funds were difficult to obtain and comprised only 26.1% of the funding program.

3.2.3.4 Analysis of Financing Costs

Because several financing sources were used for the project, the weighted average cost of the project was treated as the cost of financing the entire project. The financing costs (k) totaled 4.03%—far less than the investment profit rate of 14.94% cited in the financial evaluation report.

3.2.4 Operating Conditions and Evaluation

3.2.4.1 Forecast and Evaluation of Traffic

The actual traffic volume and engineering feasibility study projections compared in Table 3.4. The volume projected in the feasibility study report was significantly higher than actual volume.

The reasons for the gap between predicted and actual traffic flows are:

1. Premature inclusion of transfer traffic volume and induced traffic volume.
2. The additional traffic volume expected from port development has not been fully exploited.

Table 3.3 Financing Stages of Change

Sources of Funding	Feasibility Study Financing		Preliminary Design Financing		Actual Project Financing	
	Amount (Millions)	Total Investment Ratio (%)	Amount (Millions)	Total Investment Ratio (%)	Amount (Millions)	Total Investment Ratio (%)
Ministry of Communications grant	14.5	19.96	10.28	13.05	10.28	15.74
Provincial highway funds	30	41.31	8.87	11.27	8.82	13.51
City matching funds	0	0	9.02	11.45	3.35	3.6
Transportation Industry Group	0	0	1.4	1.78	1.4	3.14
Long-term loans	11.57	15.93	43.73	54.23	34.45	53.76
Treasury lending	0	0	0	0	6.5	9.95
Corporate bonds	0	0	0	0	1.5	3.3
Japanese bank loans	16.56	23.8	0	0	0	0
Special funds	0	0	6.5	8.25	0	0
Total	73.63	100	78.8	100	65.3	100

Table 3.4 Comparison of Actual and Predicted Traffic Volume A(Vehicles/Day)

Date	2003	2004	2005
Predicted traffic	15,276	15,706	16,158
Actual traffic	6,006	8,753	9,343
Error rate	60%	44%	42%

Note: Error rate = (predicted value – actual value)/max {actual value, predicted value}.

3. To protect traffic safety and extend road life, the penalties for overloaded vehicles were increased.
4. The uncertainty of regional economic development will affect traffic volume. The forecast of urban and port economic develop was too optimistic, and led to errors in predicting traffic volume.

3.2.4.2 Analysis of Vehicle Speed

According to vehicle speed statistics obtained at specific monitoring points, the average vehicle speeds for various sections are shown in Table 3.5.

As shown in the table, the average speeds over three sections of the highway range from 75 to almost 90 km/h. The Eastern section shows the highest speeds, followed by the Central section. The Western section speeds were the lowest. Since the highway opened, the average vehicle speed improved significantly, shortening travel times between cities along the route and easing congestion on other routes.

3.2.4.3 Evaluation of Structural Changes in Traffic

Figure 3.3 shows a large gap between actual and predicted traffic (particularly large truck traffic) cited in the feasibility study. The actual values for other vehicle exceeded the predictions.

3.2.4.4 Evaluation of Traffic Safety Management

Since the highway opened to traffic, safety of drivers, passengers, and workers engaged in construction and maintenance has been emphasized (Table 3.6). Five highway patrol brigades distributed near cities along the highway are responsible for patrolling and handling accidents.

According to the table, in 2003 when the highway was newly opened, both traffic volume and accident rates were low. The increased traffic volume after 2004 led to increases in accident and death rates. Research revealed that driver fatigue,

Table 3.5   Average Speed of Highways Vehicle (km/h)

Date	2003	2004	2005
Eastern section	84.4	87.6	87.4
Central section	79.9	84.7	81.3
Western section	74.2	75.9	78.3
Highway a-verage	79.5	83.7	83.3

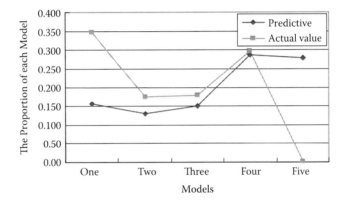

Figure 3.3   Contrast between feasibility study forecast models and actual ratio models in 2004.

speeding, and bad weather are the main causes of traffic accidents. Publicizing the dangers of driving while fatigued, a crackdown on serious traffic violations such as speeding, and better traffic management in rain, snow, fog, and other adverse weather have increased traffic safety.

3.2.5 Evaluation of Management, Support, and Service Facilities

3.2.5.1 Management

Steps taken to strengthen quality management include imposing legal responsibilities on managers, strict implementation of competitive bidding, project supervision measures, safety improvements, and increased financial management and investment control. Public bidding was fair; project supervision, safety, financial management, and investment control were effective.

Table 3.6 Highway Traffic Safety Status

	July–Dec. 2003	2004	2005	Jan.–June 2006
Number of accidents	26	161	183	86
Deaths	13	34	51	21
Injured persons	21	38	38	15
Direct economic losses (10,000)	155.8	661.7	743.1	241.5
Accident rate (Million vehicles • km)	0.102	0.216	0.23	0.189
Mortality rate (Million vehicles • km)	0.051	0.0456	0.064	0.0462

3.2.5.2 Support and Service Facilities

The highway has a centralized management system that includes two management subcenters, two conservation work areas, and ten toll stations. Routine maintenance is carried out by the highway company's staff and equipment. The service, security, and utility facilities are in operation. Highway monitoring is achieved by an intelligent, communication system.

3.3 Traffic Forecasting

3.3.1 Basis

Highway traffic flow is the number of vehicles on a section of a highway during a selected time period. Traffic flow data used for forecasts in feasibility studies is based on total quantity control, expert predictions, time series techniques, exponential smoothing, regression analysis, elasticity coefficient, growth rate, and combined modeling. However, these techniques failed to consider the effects caused by the interactions of multiple factors. We constructed a prediction model based on the regional economic system to solve this problem.

Highway traffic flow is impacted by regional economic growth trends and macroeconomic conditions. We first constructed a high-speed flow forecast of the regional economic system. Because it is a dynamic open system, external factors impact development. After analysis, significant impacts on the system were determined by considering two factors: (1) the development of ports; and (2) the construction of road networks that could divert traffic. Integrated traffic is calculated as $\alpha \times$ traffic based on trends $+ (1 - \alpha) \times$ flow based on regional transportation system + impact of port + impacts of new highways.

3.3.2 Forecasting

Ten toll stations recorded vehicles passing through 32 road sections along the Eastern, Central, and Western sections of the highway. We reviewed total traffic volume data for each section and uniformly converted it into standards for passenger cars and used the obtained data for forecasting.

3.3.2.1 Based on Trend of High-Speed Flow to Forecast

The grey model known as GM(1,1) is one of more appropriate prediction methods devised to deal with "poor information, little data." We used 4 data requirements and 3 years of the high-speed highway data collected for 2003 through 2005 as the basis to construct a GM(1,1), predict the road traffic of the first and second halves of 2006, and convert semiannual flow into annual data. Then based on 4 years of traffic data for 2003 to 2006 we constructed a GM (1,1) to predict future traffic flow.

3.3.2.2 Based on Regional Transportation System High-Speed Flow to Forecast

There are three major trunks in the regional road network. We predicted the total flow for the Eastern section, the Central section, and the Western section. We first processed monitoring data from the Ministry of Transportation, calculated standards for passenger cars; then fitted the data to predict the trend for 2005 and 2006 and used data for 2003 to 2006 to construct a GM (1,1) to predict long-term traffic flow.

3.3.2.3 Forecasting Impacts of Ports on Highway Traffic

The GM (1,1) technique was used to forecast the value of road transport of port cargo over the three sections of the of highway.

3.3.2.4 Forecasting Impacts of New Roads

During China's eleventh five-year period, three new building projects greatly impacted the Lianxu Highway. Separate analyses of growth and losses to the three routes were used to predict their impacts on the Lianxu Highway. We used the formula for calculating integrated highway traffic ($\alpha = 0.5$) and analyzed trends and regional transportation flow forecasts. The results are presented in Table 3.7.

Table 3.7 Prediction of Integrated Traffic (Vehicles/Day)

Year	Eastern Segment	Central Segment	Western Segment
2006	4,096	9,150	11,921
2007	4,612	10,641	13,833
2008	5,125	12,140	15,658
2009	5,987	13,744	17,688
2010	5,921	15,114	20,334
2011	7,009	17,065	22,895
2012	8,173	18,953	25,636
2013	9,453	20,899	28,597
2014	10,851	22,892	31,792
2015	12,000	25,155	34,853
2016	14,,013	26,943	38,907
2017	15,780	28,958	42,846
2018	17,677	30,939	47,061
2019	19,704	32,850	51,563
2020	21,655	34,884	56,148
2021	24,163	36,315	61,469
2022	26,607	37,787	66,919
2023	29,197	39,023	72,730

The analysis of the effects cumulative impacts of various factors shows large flow variations among the three sections.

3.3.2.4.1 Eastern Section: Low Starting Point, Low Growth

By comparing the traffic flows based on trending, it is evident that external factors such as new highways and port development exerted obvious impacts. Based on the prediction for integrated traffic flow, the service level will reach B in 2022. This section of the road shows the biggest gap with the feasibility report.

3.3.2.4.2 Central Section: Moderate Starting Point, Slow Growth

In the first 3 years of operation, the flow of the central section was significantly lower than the feasibility report predicted, and then the flow slowly increases. Via a comparison of traffic flow and integrated flow based on trending, we see that external factors such as new highways and port development produced no obvious impacts on this section. The flow characteristics are very similar to that in the feasibility study report.

3.3.2.4.3 Western Section: High Starting Point, Steady Growth

In the first 3 years of operation, the western section handled the most flow but did not reach the level cited in the feasibility report. The flow steadily increased after the third year. By comparing traffic flow and integrated flow based on trending, we see that external factors such as new highways and port development exerted large impacts on the road. According to the comprehensive traffic report, service level B will be achieved in 2012 and level C will be reached in 2017. After level D is reached in 2019, flow will be restricted.

Based on highway conditions and the evaluation of traffic flow, the following recommendations were made:

- Combine qualitative and quantitative analyses to improve forecast accuracy.
- Optimize the network layout and coordinate new road development.
- Construct a reasonable fee adjustment mechanism based on market factors.

3.4 Financial and Economic Evaluation

3.4.1 Financial Evaluation

3.4.1.1 Main Parameters

1. Benchmark discount rate is the minimum rate of return a project should achieve; the rate is 7% for this project.
2. Evaluation period: construction period (1997–2003) plus operating period (2004–2023) equals 27 years.
3. Rate: Sales tax rates on roll revenue were 5% before June 2005 and 3% thereafter. The rate for troubleshooting, repair and maintenance revenue, and food and beverage income is 5%. The urban maintenance construction rate is 7% of sales tax. The education surcharge is 4% of sales. Corporate income tax is 33% of profits.
4. Depreciation: The fixed assets of the project are subject to various depreciation methods (Table 3.8).
5. Reserve fund: 10% of profit after tax are drawn as statutory surplus reserves annually.
6. Statutory communal contribution: 5% of profits annually.

Table 3.8 Depreciation of Fixed Assets

Asset Class	Original Asset Value (10,000)	Estimated Residual Value (%)	Depreciation Period (Year)	Depreciation Rate (%)
Highways and structures	579,458	–	–	Traffic flow per year and ttotal estimated traffic of operation period
Safety facilities	25,352	3	10	9.70
Three systems	10,137	3	8	13.13
Mechanical equipment	2,164	3	8	13.13
Vehicles	1,665	3	8	13.13
Buildings	26,317	3	30	3.23
Other	7,586	3	5	19.40
Total	652,678	–	–	–

3.4.1.2 Revenue and Costs

The costs of the project were classified as: (1) project construction costs, operating expenses during the construction period, the fixed asset investments, and costs of trial operation; and (2) the costs of operating including collection, conservation, management fees, other operating expenses, interest, depreciation of fixed assets, and amortization of test and overhaul costs. Project revenues included tolls and other business income. In addition to business taxes, deductions from income included sales tax (VAT), urban maintenance and construction tax, and education surtax.

3.4.1.3 Financial Evaluation

3.4.1.3.1 Project Profitability Analysis

Increases in gross profits during the project evaluation were projected at 18,763.46 million yuan. Based on a tax rate of 33%; the corporate income tax will be 6,191.94 billion yuan. Undistributed profit should be 10,673.86 million yuan by deducting 10% of the reserve fund and 5% for communal purposes after income tax according to the law. The operating profit margin was 64.29%, return on investment was 14.37%, and capital profit margin was 43.13%.

3.4.1.3.2 Cash Flow Analysis

Cash flow analysis during evaluation considered financial net present value (FNPV), financial internal rate of return (FIRR), financial benefit:cost ratio (FBCR), and financial investment payback period (Pt). These four indices were used to evaluate cash flows as shown in Table 3.9.

Analysis of single-factor changes was employed to investigate all cost indicators (both operating and overhaul costs) raised or lowered by 10 or 20% and operating income (based on traffic flow and fee changes) raised or lowered by 10 or 20% (Table 3.10).

As shown in the table, the project exhibited strong resistance to single-factor changes. When the traffic flow reduced by 20% or fees increased by 20%, the internal rate of investment (before tax) and the operation's own funds remained in the normal range. By comparison, traffic flow influenced the project financial evaluation more greatly. Traffic flow reduced by 20% and all investment (after tax) rates of return were lower than the financial benchmark discount rate; net present value was also negative. To ensure traffic flow sustainability, rapid growth, is one effective strategy.

3.4.1.3.3 Analysis of Two-Factor Changes

Two-factor analysis was used to investigate changes in key financial indicators when traffic flow and the fees changed at the same time (Table 3.11).

Note the strong resistance to a two-factor change. When the traffic flow is reduced by 20% and the fee is increased by 20%, the internal rate of investment (before tax) and the operation's own funds exceeded the benchmark rate of return and the payback period was less than calculated. Thus, the project was able to recover its investments during the evaluation period. However, for a 20% reduction in traffic flow, total investment (after tax) return is less than the financial benchmark discount rate. Again, rapid growth is the key to sustainability.

Table 3.9 Cash Flows during Evaluation

Index	Unit	All Investments (After Tax)	All Investments (Before Tax)	Own Funds	Remarks
FIRR	Percent	7.58	9.21	10.50	
FNPV	Million yuan	422.67	1821.90	1472.37	$I_c = 7.00\%$
Pt	Year	19.20	18.18	20.31	Including construction period
FBCR	--	1.05	1.31	1.22	$iI_c = 7.00\%$

Table 3.10 Sensitivity Analysis of Single-Factor Changes

Index		Changes of Traffic Flow				Changes of Cost			
		+20%	+10%	-10%	-20%	+20%	+10%	-10%	-20%
Total investment before tax	FIRR (%)	10.50	9.88	8.48	7.70	9.21	9.04	9.12	9.29
	FNPV (million yuan)(m)	3118	2470	1174	526	1822	1673	1748	1896
	Pt (year)	17.17	17.64	18.77	19.46	18.18	18.35	18.27	18.10
	FBCR	1.51	1.41	1.20	1.10	1.31	1.28	1.29	1.32
Total investment after tax	FIRR (%)	8.62	8.12	7.01	6.40	7.58	7.39	7.48	7.68
	FNPV (million yuan) m	1261	841	691	-407	423	279	351	495
	Pt (year)	18.33	18.74	19.73	20.33	19.20	19.43	19.32	19.09
	FBCR	1.15	1.11	1.00	0.94	1.05	1.04	1.05	1.06
Own Funds	FIRR (%)	11.59	11.07	9.88	9.20	10.50	10.21	10.35	10.64
	FNPV (million yuan) (mi	2104	1790	1154	835	1472	1342	1407	1538
	Pt (year)	19.67	19.96	20.51	20.58	20.31	20.51	20.41	20.21
	FBCR	1.28	1.25	1.18	1.14	1.22	1.20	1.21	1.23

Table 3.11 Sensitivity Analysis of Two-Factor Changes

Index	Change of Traffic Flow Change of Fee	+ 20%	+ 10%	0%	−10%	−20%
FIRR (%)	+ 20%	10.34	9.72	9.04	8.31	7.52
Total investment before tax	+ 10%	10.42	9.80	9.12	8.40	7.61
	0%	10.50	9.88	9.21	8.48	7.70
	−10%	10.58	9.95	9.29	8.57	7.78
	−20%	10.65	10.03	9.37	8.65	7.87
FIRR (%)	+ 20%	8.44	7.93	7.39	6.81	6.20
Total investment after tax	+ 10%	8.53	8.02	7.48	6.91	6.30
	0%	8.62	8.12	7.58	7.01	6.40
	−10%	8.72	8.21	7.68	7.11	6.50
	−20%	8.81	8.30	7.77	7.20	6.60
FIRR (%)	+ 20%	11.31	10.79	10.21	9.59	8.90
Own funds	+ 10%	11.45	10.93	10.35	9.73	9.05
	0%	11.59	11.07	10.50	9.88	9.20
	−10%	11.72	11.20	10.64	10.02	9.35
	−20%	11.86	11.34	10.77	10.17	9.49
PtT (year)	+ 20%	17.33	17.83	18.35	18.97	19.68
Total investment before tax	+ 10%	17.25	17.73	18.27	18.87	19.57
	0%	17.17	17.64	18.18	18.77	19.46
	−10%	17.09	17.55	18.10	18.67	19.36
	−20%	17.01	17.47	18.01	18.57	19.25
PtT (year)	+ 20%	18.54	18.97	19.43	19.98	20.60
Total investment after tax	+ 10%	18.43	18.85	19.32	19.85	20.46
	0%	18.33	18.74	19.20	19.73	20.33
	−10%	18.23	18.63	19.09	19.60	20.19
	−20%	18.13	18.53	18.98	19.48	20.07

Table 3.11 (Continued)   Sensitivity Analysis of Two-Factor Changes

Index	Change of Traffic Flow Change of Fee	+ 20%	+ 10%	0%	−10%	−20%
PtT (year)	+ 20%	19.86	20.16	20.51	20.92	21.22
Inherent funds	+ 10%	19.77	20.06	20.41	20.70	21.04
	0%	19.67	19.96	20.31	20.51	20.58
	−10%	19.58	19.85	20.21	20.33	20.01
	−20%	19.50	19.74	20.11	20.18	19.87

3.4.2 Economic Evaluation

1. Evaluation parameters:
   a. Evaluation period: Construction period plus 20 years of operation.
   b. Social discount rate: The shadow price of capital, reflecting the opportunity cost of capital and time value of money; the rate was set at 8% postevaluation.
   c. Shadow wages: Based on provincial labor statistics for skilled labor on highway construction projects, the shadow wage conversion factor was set at 1 for nominal wages.
   d. Residual value is 50% of construction cost treated as a negative economic cost.
2. Economic costs (construction, collection, maintenance, management, major changes, and other costs). Construction costs include installation, equipment purchases, and other costs. The costs of land, labor and materials were adjusted to the shadow of the main costs. Other costs were taxes, utility fees, interest on loans from domestic banks, and other items. Road maintenance and overhaul costs were simplified by using an adjustment factor of 0.959. Charge fees included station wages, management, and maintenance expenses. Transfer payments ceased in 2004. The annual adjustment coefficient is 0.83; half the construction cost is taken as residual value.
3. Economic benefits (direct and indirect) measure contributions to the national economy. Only direct benefits were quantitatively calculated in the postevaluation; indirect benefits were examined by qualitative analysis and include (1) reduction of transportation costs; (2) transportation time savings; and (3) reduction of accident losses.

4. Evaluation results: Key indicators used to measure the feasibility and economic benefits of the project were economic net present value (ENPV), financial internal rate of return (FIRR), economic benefit:cost ratio (EBCR), and economic payback period (EN). See Table 3.12.

5. Sensitivity analysis: Uncertainty in the economic evaluation of all factors considers costs and benefits by analyzing the cost effectiveness of changes in the economic evaluation to determine the degree of change. As shown in Table 3.13, the project can face risk. At worst, when costs increased by 20%, benefits reduced by 20%. The internal rate of return could be kept at 10.31%—above the 8% social discount rate. From the perspective of national economic evaluation, the project's operations are reasonable and correct.

3.4.3 Comparisons of Feasibility Study and Postevaluation

3.4.3.1 Comparison of Financial Benefits

The operation period was set at 20 years in both the feasibility study and postevaluation. The benchmark rate of return was set at 4.17% based on current prices and interest rates. Because of interest rate adjustments, the benchmark rate of return was conservatively reset at 7% in the postevaluation. The indicators are compared in Table 3.14.

The financial results from analyzing the feasibility and postevaluation stages indicate the project yielded good financial benefit. However, the indicators (particularly net present value and benefit:cost ratio) generally declined during postevaluation evaluation due to:

1. Differences in project cost. During construction, construction material costs remained stable and did not require expenditures of reserve funds. Interest rate reductions saved 10.14% resulting in a project cost of 6,526,78 million yuan.

Table 3.12 Indices of Economic Evaluation

Number	Index	Unit	Outcome	Remarks
1	Financial internal return rate (FIRR)	Percent	13.88	
2	Economic net present value (ENPV)	Million 10,000yuan	5795.62	$i_s = 8.00\%$
3	Investment payback (Pt)	Year	19.15	Including construction period
4	Economic benefit:cost ratio (EBCR)	–	3.21	$i_s = 8.00\%$

Table 3.13 Sensitivity Analysis

Index	Change of Traffic Flow / Change of Cost	+20%	+10%	0%	−10%	−20%
FIRR (%)	+20%	13.88	13.32	11.71	11.04	10.31
	+10%	13.46	13.88	13.26	11.59	10.85
	0%	14.09	13.51	13.88	13.20	11.45
	−10%	14.81	14.22	13.58	13.88	13.12
	−20%	15.63	15.02	14.37	13.66	13.88
ENPV (million yuan) (10,000)	+20%	6954.74	5895.26	4835.77	3776.28	2716.79
	+10%	7434.67	6375.18	5315.69	4256.21	3196.72
	0%	7914.59	6855.11	5795.62	4736.13	3676.64
	−10%	8394.52	7335.03	6275.55	5216.06	4156.57
	−20%	8874.45	7814.96	6755.47	5695.98	4636.50
EN (year)	+20%	19.14	19.48	19.90	20.31	20.79
	+10%	18.78	19.14	19.52	19.98	20.43
	0%	18.37	18.74	19.14	19.56	20.06
	−10%	17.96	18.30	18.70	19.14	19.61
	−20%	17.44	17.81	18.21	18.64	19.14
EBCR	+20%	1.58	1.45	1.31	1.18	1.05
	+10%	1.72	1.58	1.43	1.29	1.15
	0%	1.89	1.74	1.58	1.42	1.26
	−10%	3.10	1.93	1.75	1.58	1.40
	−20%	3.37	3.17	1.97	1.78	1.58

2. Traffic differences. Actual traffic volume was lower than the flow forecast during construction. The decreased traffic negatively affected the financial benefits.
3. Fee adjustments. The highway fees increased over those projected during the construction stage. The increased fees produced positive financial effects.

Table 3.14 Comparison of Financial Efficiency Indicators

Index		Feasibility Stage	Postevaluation	Change Value	Percent Change
Pre-tax investment	FIRR (%)	10.9	9.21	−1.69	−15.54
	NPV (million yuan)V (10,000)	7668.68	1821.90	−5846.78	−76.24
	PtT (year) (year)	18.00	18.18	0.18	1.00
	FBCR	3.26	1.31	−0.95	−43.22
After-tax investment	FIRR (%)	8.80	7.58	−1.22	−13.86
	NPV (million yuan) (10,000)	4605.12	422.67	−4182.45	−90.82
	Pt (year) (year)	20.00	19.20	−0.80	−3.98
	FBCR	1.5	1.05	−0.45	−29.67
Own funds	FIRR (%)	9.98	10.50	0.52	5.16
	NPV (million yuan) (10,000)	4330.05	1472.37	−2857.68	−66.00
	Pt (year) ()	20.00	20.31	0.31	1.53
	FBCR	1.46	1.22	−0.24	−16.33

3.4.3.2 Comparative Analysis of Economic Benefits

The evaluation considered construction time plus an operation period of 20 years. The discount rate for the community was set at 12% in the feasibility study. Because of national economic changes, the rate was set at 8% in the postevaluation. For ease of comparison, we temporarily set the discount rate at 12% for calculations of net present value and costs. The comparison is shown in Table 3.15.

Every indicator was better during feasibility than postevaluation, especially ENPV and EBCR, mainly because the traffic volume predicted during construction

Table 3.15 Comparison of Economic Indicators during Feasibility Stage and after Evaluation

Index	Feasibility Stage	Postevaluation	Change Value	Percent Change	Remarks
FIRR (%)I	15.36	13.88	−3.48	−16.13	
ENPV (million 10,000 yuan)	1852.47	605.68	−1246.79	−67.30	12.00%
Pt (year)(N;)	20	19.14	−0.86	−4.29	Construction period included
E(BCR)	1.46	1.10	−0.31	−21.51	12.00%

was significantly higher than actual. Traffic volume is key to this economic evaluation because it is a direct effect.

3.5 Assessment of Environmental and Social Impacts

3.5.1 Environmental Impacts

3.5.1.1 Evaluation of Environmental Management

3.5.1.1.1 Preconstruction Environmental Management

Even before the project started, the provincial and municipal highway authorities considered environmental and ecological protection along the highway route. This ensured consistent environmental management.

3.5.1.1.2 Environmental Management during Construction

During construction, an organization was formed to handle environmental protection matters. It was responsible for setting and implementing environmental policies and regulations, routine environmental monitoring, and related activities. The route was selected to avoid villages. Local water and soil conservation measures alone the route were implemented and a drainage system was built.

3.5.1.1.3 Environmental Management throughout Project

All parties involved in the project prepared work plans to ensure environmental protection, supervised activities, trained staff, and procured appropriate equipment.

3.5.1.2 Implementation of Environmental Protection Measures

The highway route was designed to avoid villages. The local water conservation measures were reviewed and improved. A new drainage system was built and green areas were implemented and protected. The highway environmental protection office supervised the implementation of the environmental protection measures. Companies involved in construction observed the relevant requirements. The highway project included highly efficient systems for managing wastewater treatment and solid waste recycling. All these efforts produced environmental benefits.

3.5.1.3 Conclusions of Survey

3.5.1.3.1 Environmental Impacts

The investigation covered 11 toll stations, 6 service areas, and 3 managing and monitoring centers.

3.4.1.3.2 Main Conclusions

At present, borrow pits on the sides of the highway are mostly used as fish ponds by local residents; they are not connected with local water systems. Sewage discharged from the service area treatment facilities is processed by treatment facilities and released into nearby irrigation canals and ditches .The organic wastewater from filling and repair stations fosters the growth of poisonous organisms and poses risks to the surrounding ecology. The problems generated by the mixing plants have not been solved. The project plans included noise restrictions. Some residents had to be persuaded to relocate. Much of the highway was built on farm land.

3.5.2 Social Impacts

3.5.2.1 Division of Area of Coverage

The areas of the Lianxu Highway were classified as direct and indirect influences in the feasibility report (Figure 3.4). The highway passes through the direct influence areas of Lianyungang City, Donghai County, Xuzhou City, Xinyi City, Pizhou City, and Tongshan County. Indirect influence areas included 15 cities or counties in Jiangsu and Anhui Provinces and Tongshan County.

3.5.2.2 Economic Development Correlation of Highway and Line Side Areas

The economies of the side areas grew quickly. From 2000 to 2004, the gross domestic product (GDP) growth rate (average = 13.2%) was higher than the rates in other areas (7.9%) (Figure 3.5). After the completion of the project, the economic growth

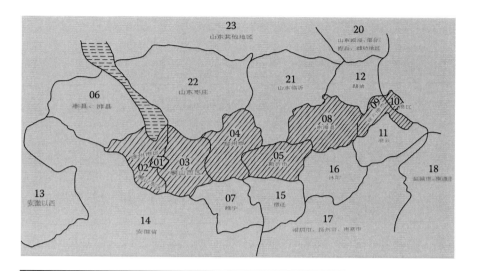

Figure 3.4   Project impact zone divisions.

rate of the side areas accelerated; transportation and economic development in the influence area also increased.

3.5.3 Economic Development

1. Direct contributions of the project to the economies along the highway were measured:

   a. Economic growth of areas along the project: The contributions to regional economic growth along the project (freight volume and GDP elasticity coefficient) were used to calculate the highway's contribution to direct influence areas and GDP growth:

$$E = \frac{\Delta X / X}{\Delta Y / Y}$$

where E is the elasticity coefficient, X is the freight volume, ΔX represents the growth of goods and passengers transported, and ΔY is the GDP growth resulting from highway operation. We calculated the freight volume on the highway with and without antithesis. According to the original freight volume data for 1999 through 2003, a grey forecasting model predicted the freight volume without the highway. We then compared the projected and actual freight volumes and calculated the difference, that is ΔX. The elasticity coefficient E was based on feasibility study statistics; consequently, the contribution quantities and contribution rates for directly affected zones could be calculated. In 2004, the total contribution to GDP in directly affected zones was 1.488 billion yuan; the corresponding general contribution rate was 10.33%. Clearly, the highway has produced immense outputs and led to rapid development of the economies in line side areas.

b. Other contributions: Efficient highway transportation systems expand markets and industrial activities. The Lianxu Highway linked the cities along its route and improved road transportation throughout the area serviced.

2. Influence on social and economic development in the region along the line:
   a. Improved investment potential, accelerate the development of new industries and improving the local economies.
   b. Promotion of regional cooperation and economic development.
   c. Protection of land resources.
   d. Accelerating the development and ensuring reasonable distribution of urban population.
   e. Creation of employment opportunities.
   f. Cultivation of managerial talent to enhance technological progress of the national highway system.
3. Adverse effects:
   a. Ineffective use of land resources.
   b. Contamination of environment.
   c. Damage to ecological balance.

3.5.4 Macroeconomic Impact Analysis

3.5.4.1 Impact on Total Economy

Inputs and outputs were analyzed. The total efficiency of investment equals the sum of investment efficiency and investment multiplier efficiency. The investment efficient was calculated as 5663.2818 million yuan. The investment multiplier efficiency was 11326.5636 million yuan; thus, the total investment efficiency was 16989.8454 million yuan or 3.6 times the inputs. According to formula for computing elasticity coefficient, the contribution (passenger and cargo transportation) to the GDPs of three provinces were calculated. Table 3.16 shows the results.

Table 3.16 Highway Contributions to GDP in Three Provinces in 2004

Project	Province 1	Province 2	Province 3
Contribution volume (billion yuan))	6.77	5.89	1.328
Contribution rate	3.30%	1.93%	1.45%

3.5.4.2 Effects on Economic Structure

The planning, building, and operation of the highway contributed enormously to a number of industries including construction, transportation, and tourism and other services.

3.5.4.3 Effects on Environment and Society

The two aspects of effects on environment and society are the occupation of land by investment unit and employment effects. The investment unit was calculated as 10 square meters per 10,000 yuan. The related coefficient of the sum of freight traffic of the highway and the sum of the quantity of employment is 94.33%. This means that the highway contributed to increased employment in the surrounding areas.

3.5.4.4 Local Compatibility Analysis

The area has abundant labor resources. National investments represent less than 0.01% of the national revenues; in some cities, the investment level is less than 0.4% of expenditures. The nation, provinces, and cities were capable of undertaking construction of the project.

3.6 Sustainability Evaluation of Project Objective

3.6.1 Effects of External Conditions

3.6.1.1 Socioeconomic Development

Policy and support by the province ensured sustainability of the project objective. The booming domestic commodity trade further ensures project sustainability.

3.6.1.2 Highway Network Development

China's transportation infrastructure is vital. The highway network promotes the development of the regional economies and helps guarantee the sustainable development of this highway.

3.6.1.3 Transportation Development

Highway passenger and freight traffic continues to grow. The rapid growth in port traffic has increased highway traffic, thus aiding sustainable development.

3.6.1.4 Management System

The enterprise management system utilized on the highway continues to produce positive and profound impacts on sustainable development.

3.6.1.5 Policies and Regulations

Highway law, policies, and regulations provide a legal framework for promoting sustainable and healthy development.

3.6.1.6 Supporting Facilities

Comprehensive toll, security, and other services provide a good foundation for attracting more traffic and increasing income to ensure sustainability of the project development objectives.

3.6.2 Effects of Internal Conditions

3.6.2.1 Operating Mechanism

A strict and sound management system and start-of-the-art equipment have played positive roles in the sustainable development of the highway.

3.6.2.2 Internal Management

A set of unique, efficient management and operation mechanisms provides internal security mechanisms to ensure the sustainable development objectives of the project are met.

3.6.2.3 Service Status

The highway route and placement of facilities are well designed. Configurations, signage, and complete service facilities are complete and function well. The operator maintains the road and provides quality services to customers to sustain the project development objectives.

3.6.2.4 Impacts of Highway Tolls

The steady growth of traffic volume and toll rates will increase the revenue of the highway and promote sustainable development.

3.6.2.5 Impacts of Operation Conditions

Operation quality directly affects sustainability. Since the highway opened, the operating company continued to strengthen management, and operating income increased significantly.

3.6.2.6 Impacts of Construction Quality

Excellent construction quality laid a solid foundation for ensuring the sustainability of the project objectives.

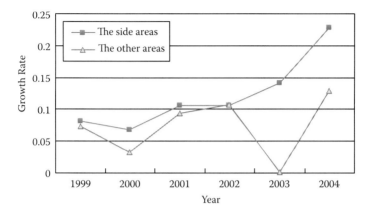

Figure 3.5   GDP growth rate in the side areas and other areas.

3.6.3 Comprehensive Evaluation of Sustainability

3.6.3.1 Evaluation Index

Based on operations data and ongoing evaluation of project objectives, a comprehensive evaluation index system for the sustainability of the project objectives was devised as shown in Figure 3.6.

3.6.3.2 Determination of Index Weight

We utilized the analytic hierarchy process (AHP) to determine the weights of indices and questionnaires directed to experts, company managers, and engineers to compare the importance of each index. Applying investigation and statistical results via a hierarchy judgment matrix, we determined index weights (Table 3.17).

3.6.3.3 Conclusion

The target sustainability comprehensive evaluation coefficients shown in Table 3.18 and Figure 3.7 show an upward trend. In the first 3 years after completion (2003–2005), the target sustainability coefficient increased slowly; the years from 2007 to 2023 show an upward trend that should increase gradually.

3.6.4 Means for Realizing Sustainability

1. Effectively maintain the highway.
2. Improve operation and management capabilities and establish a customer-oriented business philosophy.

Table 3.17 TIndex Weights of Sustainability Objectives

Level Indicator	Weight	Secondary Indicator	Weight
Index of service capabilities	0.3709	Vehicle saturation	0.5
		Quality	0.5
Index of operational sustainability	0.3952	Operating income	0.4073
		Management capacity	0.5927
Index of social and economic benefits	0.2339	Contribution ability for per capital GDP of areas along line	0.6082
		Contribution ability for living of areas along line	0.3918

3. Position service areas reasonably and make full use of them.
4. Protect the ecology and environment along the highway.

3.7 Problems and Recommendations

3.7.1 Problems

3.7.1.1 Defective Analysis of Effects on Transfer, Induced Traffic Volume, and Gap between Forecast and Actual Value

Experience indicates that transfers and increases of traffic volume occur gradually 2 to 3 years after the completion of construction rather than immediately. The feasibility report prematurely predicted additions to traffic volume, resulting in errors in traffic forecasts.

3.7.1.2 Inadequate Design Based on Engineering Survey of Local Roads: Increased Design Costs and Impacts on Progress

The geologic drilling density is rather low. The projected distributions of local soft soil and liquefied soil for the project were inaccurately reflected in the engineering exploration report. This created geological problems such as subgrade cracking and slipping. The design had to be modified; this increased costs and impeded progress.

Table 3.18   Twenty-Year Sustainability Projection for Highway Project

Year	Service Capability Index	Operational Sustainability Index	Social and Economic Benefit Sustainability Index	Project Objective Sustainability Index
2004	1.00	1.00	1.00	1.00
2005	0.96	1.03	1.34	1.08
2006	1.00	0.87	1.12	0.98
2007	0.92	1.00	1.14	1.00
2008	0.86	1.12	1.21	1.05
2009	0.81	1.27	1.32	1.11
2010	0.76	1.50	1.45	1.21
2011	0.71	1.70	1.59	1.31
2012	0.68	1.91	1.73	1.41
2013	0.64	3.12	1.89	1.52
2014	0.63	3.34	3.04	1.64
2015	0.59	3.81	3.19	1.84
2016	0.56	3.09	3.34	1.98
2017	0.52	3.36	3.49	3.10
2018	0.49	3.63	3.62	3.23
2019	0.46	3.91	3.75	3.36
2020	0.44	4.61	3.87	3.66
2021	0.41	4.92	3.99	3.80
2022	0.38	5.22	3.09	3.93
2023	0.35	5.53	3.19	3.06

3.7.1.3 Inadequate Funding Programs and Municipal Matching Requirements

Government funds covered only 26.1% of project costs. The funding projections were not reasonable. To compensate for the lack of funds, diversified sources were utilized. This increased the cost and risks of financing.

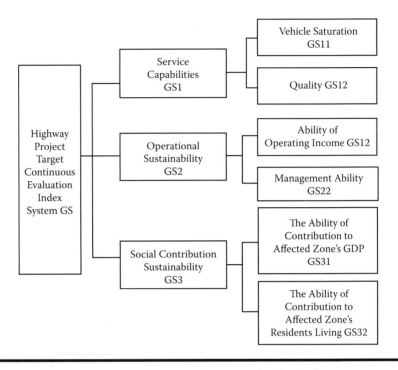

Figure 3.6   Highway project target continuous evaluation index system.

3.7.1.4 Unreasonable Standards for Service Facilities: Wastes of Resources

Business costs for the service areas from January 2003 to October 2005 were slightly larger than business revenue. As a result, the functional and economic benefits of the service area were not fully realized. Some standards for service facility construction were unnecessarily strict. As a results costs exceeded budget and led to waste of resources.

3.7.1.5 Preliminary Test Section

Construction started before the authorities agreed. A 5-km test section was constructed and increased the risk of the project.

3.7.1.6 Inadequate Environmental Protection and Pollution

Effective measures for dealing with environmental issues were not designed effectively. Failure to install appropriate pits for sewage treatment in services, failure to

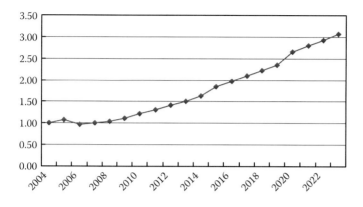

Figure 3.7   Target sustainability comprehensive evaluation coefficient of highway.

restore vegetation in open areas, and lack of noise control increased environmental pollution along the route.

3.7.1.7 Land Waste Caused by Defective Selection Programs for Local Roads

Local roads and embankments were not adequate and effective roadbed height was difficult to determine.

3.7.2 Recommendations

Overall, the results of highway operation starting from construction through full operation were satisfactory. Analyzing financial, economic, and social benefits indicated that most expect goals of the project were achieved. In response to problems that arose during construction and operation, some recommendations and solutions were proposed:

1. Consider impacts of the transfer and induced traffic and improve traffic flow prediction accuracy. Highway traffic growth depends on factors such as location, natural resources, social economy, population, vehicle ownership, per capita income, land use, road service features, supply efficiency, and development of other modes of transport. The feasibility report forecast large deviations of the volumes of transfer and induced traffic. Seeking expert advice, identifying critical factors, and considering the highway's role in the national network are important aspects of feasibility studies.
2. Strictly follow construction specifications, save resources, and prevent risks. Accurate specifications guarantee quality of a finished product. Failing to

meet specifications leads to substandard construction, increases costs, delays progress, affects local government participation, impacts management control, and wastes resources and time. Strict implementation of construction specifications, effective management, and thorough quality control can essentially eliminate risks, repairs, and rework.

3. Design reasonable funding programs and eliminate financial risks. For large-scale infrastructure projects, funding sources should be examined comprehensively to avoid funding gaps during construction. It is wise to consider a wide range of funding sources and methods to stimulate domestic and foreign investment in road and bridge projects.

4. Use embankment programs to save and protect land resources. Factors to consider in selecting an appropriate embankment program include geological conditions, road grade standards and use requirements, field conditions, and potential for environmental impacts. Theoretical and quantitative analysis can provide reliable bases for program designs.

5. Service facilities should be constructed according to users' needs to avoid blind comparisons and image projects. A design should meet the needs of the service area and traffic flow without wasting resources.

6. Devise environmental protection measures and strengthen them after project completion. Environmental protection should cover the entire life cycle of a project, from construction on and it requires more attention when a project is fully operational. Issues such as technical or financial problems arising from environmental restoration should be negotiated with local governments before a project starts and the details should be clearly explained in engineering documents.

7. Coordinate traffic, maintenance, operations, management, and other aspects of a project to meet sustainability targets. Adequate technology should be used to coordinate traffic flow. A flexible and reasonable standard fee system should be devised. Road maintenance and environmental protection measures should be monitored consistently and improved when possible to maintain sustainability. Environmentally friendly materials should be used; if possible, farm land should not be used for highway construction. Sustainability and effective resource use are critical factors even in the planning stages.

Chapter 4

Efficiency Evaluations of Scientific and Technological Activities

4.1 Introduction

After decades of development, science and technology in China continue to thrive. Technological competition and economic power show remarkable rises. China's science and technology inputs are increasing, and its share of gross domestic product (GDP) is also rising. In 2008, the total national expenditures for research and development totaled 4616.0 billion RMB—90.58 billion RMB—more than 24.4% growth over 2007. Science and technology talent has increased rapidly; 22 of every 10,000 workers are employed in research and development (R&D) and that number continues to increase.

Although the inputs and outputs of science and technology are growing, the low efficiency in those areas represents a serious problem. According to statistics generated since 1990, the per capita funds for science and technology increased from 20,300 RMB in 1990 to 235,000 RMB in 2008—an 11.58-fold increase. However, the outputs of R&D funds per million RMB for R&D continue to decrease. Numbers of publications and patent licenses per million RMB of R&D funds at home and abroad have declined. If inflation is taken into account, the increase of funds per capita and the decreases of outputs are still obvious. The number of scientific and technological achievements is decreasing more drastically because of government-adjusted science and technology-praising policies.

The low efficiency of science and technology activities restricts development and wastes precious resources. For this reason, we evaluate the efficiency of science and technology activities in this chapter and reveal key factors affecting efficiency. We also provide suggestions to promote the efficiency of science and technology activities in China.

4.2 Allocation Structure and Use Efficiency Analysis of Research Expenditures

Scientific and technological accomplishment is gradually becoming the symbol of comprehensive strength of a nation. The intensity of investment for science and technology relates directly to scientific and technological advances and the efficient use of funds. This chapter systematically studies allocation structure and use efficiency from 1999 to 2007 and compares several efficiency factors.

In 1999, the R&D investment was 67.891 billion RMB. Over the next 8 years, the total increased 4.46 times to 371.02 billion RMB. This growth rate indicates that science and technology investment exceeds GDP. Thus, our society values scientific investment. The concept that "investment for science and technology is a high-profit productive investment" is widely accepted throughout China. The nation's scientific expenditures reached 170.36 billion RMB—3.6 times the level in 1999, exceeding GDP growth and ranking as the greatest expenditure in the past 10 years.

In 1999, per capita funds for science and technology were 116 RMB and 54 RMB for R&D in China. In 2007, per capita funds reached to 537.28 RMB and 280.79 RMB, representing growth rates of 363 and 420%, respectively—higher than the growth of GDP. Compared with other countries, particularly developed countries, China needs to improve. In 1996, per capita funds for R&D equaled $720 in the United States and £246 in Great Britain. We must make full use of somewhat limited investment to develop our scientific and technological enterprises and try to improve our scientific and economic strength and competitiveness. We must ensure optimum resource allocation constantly increase the use efficiency of funds for science and technology.

4.2.1 Allocation of Funds for Science and Technology in China

The total funding for science and technology in 2007 was raised to 769.52 billion RMB, representing growth of 427% compared with 146.06 billion RMB in 1999. The government appropriated 170.36 billion RMB, an increase of 260%; self-raised funds were 518.95 billion RMB, an increase of 596%; bank loans were 38.43 billion RMB, an increase of 198%. The bank loans represent less than half of the national average growth rate.

In 2007, the funds raised by large and medium-sized enterprises accounted for 56.4% of the country's total funds for science and technology compared with 45.56% in 1999; this represents growth of about 11 percentage points. The markets certainly affected allocations of scientific and technological resources and economic and social reform in China increased the dominance of non-government enterprises. Based on government expenditures, funds granted to scientific research institutions and large and medium-sized enterprises are trending downward; amounts granted to colleges and universities are trending up, meeting the requirements of optimizing allocations of scientific and technological funds and consistent with my suggestions in 1999.

Government-appropriated funds for scientific research institutions in 2007 accounted for 82.39% of their funds compared with 62.9% in 1999, representing a growth of about 20 percentage points. This shows that the dependence of scientific research institutions on government-appropriated funds has not changed despite reform efforts.

As shown in Tables 4.1 and 4.2, the number of personnel engaging in scientific and technological activities is trending downward, and the number of engineers

Table 4.1 Per Capita Funds of Personnel Engaging in Scientific and Technological Activities in 1999

Item	Unit	Total	Scientific Research Institutions	Colleges and Universities	Large and Medium-Sized Enterprises
Personnel engaging in scientific and technological activities	Thousand	2906.0	553.0	342.0	1453.0
Per capita funds	Thousand RMB	50.3	98.1	30.1	45.8
Appropriated funds per capita	Thousand RMB	16.3	61.7	14.4	3.4
Scientists and engineers	Thousand	1595.0	342.0	329.0	668.0
Per capita funds	Thousand RMB	91.6	158.6	31.2	99.6
Appropriated funds per capita	Thousand RMB	29.7	99.8	15.0	7.4

Table 4.2 Per Capita Funds of Personnel Engaging in Scientific and Technological Activities in 2007

Item	Unit	Total	Scientific Research Institutions	Colleges and Universities	Large and Medium-Sized Enterprises
Personnel engaging in scientific and technological activities	Thousand	4544.0	478.0	542.0	2202.0
Per capita funds	Thousand RMB	169.3	264.5	113	195.8
Appropriated funds per capita	Thousand RMB	37.5	217.9	63.7	6.6
Scientists and engineers	Thousand	3129.9	356.0	460.0	1401.0
Per capita funds	Thousand RMB	245.9	354.2	133.3	308.0
Appropriated funds per capita	Thousand RMB	54.4	291.8	75.1	10.3

is trending upward. The personnel structures of scientific research institutions are becoming more reasonable and the personnel quality continues to improve. Judging by the per capita expenditures for personnel engaging in scientific and technological activities, the expenditures for personnel from colleges and universities are the lowest. Despite growth of about 6 percentage points in government-appropriated funds per capita in 2007, reaching 63,700 RMB, it only accounted for 29.23% of government-appropriated funds per capita for personnel from scientific research institutions. If calculated on the basis of scientists and engineers engaging in scientific and technological activities, per capita funds for scientific research institutions and large and medium-sized enterprises were 2.65 and 2.3 times, respectively, the funds for colleges and universities. This is an obvious decrease in relation to the 5 and 3, respectively, calculated for 1999. This shows that the government-appropriated funds are increasingly going to academic institutions.

The output of scientific and technological activities can reflect the use efficiency of funds for such activities to an extent. The use efficiency of funds for science and technology of all kinds is based on three indices: (1) theses published in Chinese scientific and technological magazines; (2) authorized patents; and (3) important achievements.

Judging by the scientific and technological outputs (Tables 4.3 and 4.4), the large and medium-sized enterprises dominate in terms of authorized patents and important achievements that indicate scientific and technological innovation and application. The number of theses published by colleges and universities exceeds 60% of the total for 2007. This indicates that academic institutions are making great progress in basic research, thus providing bases for scientific and technological innovation in China.

Judging by the output of funds for science and technology (Tables 4.5 through 4.8), the numbers of theses and important achievements per million RMB for science and technology and R&D in 2007 were lower than in 1999 and the outputs of scientific research institutions and large and medium-sized enterprises per million RMB were lower than in 1999. Based on the outputs of funds for science and technology in 2007, the outputs of colleges and universities per million yuan were many times higher than outputs of scientific research institutions and large and medium-sized enterprises.

Table 4.3   Scientific and Technological Activity Outputs in 1999

Item	Total	Scientific Research Institutions	Colleges and Universities	Large and Medium-Sized Enterprises
Theses	133,341	25,751	86,921	8,745
Authorized patents	23,918	1,829	960	20,229
Important achievements	22,178	6,863	7,336	7,979

Table 4.4   Scientific and Technological Activity Outputs in 2007

Item	Total	Scientific Research Institutions	Colleges and Universities	Large and Medium-Sized Enterprises
Theses	463,122	47,189	305,788	14,785
Authorized patents	351,782	4,196	14,111	43,652
Important achievements	34,170	6,263	7,592	12,220

Table 4.5 Output of Scientific and Technological Funds per Million RMB in 1999

Item	Whole Country	Scientific Research Institutions	Colleges and Universities	Large and Medium-Sized Enterprises
Theses	1.03	0.49	10.23	0.16
Authorized patents	0.19	0.03	0.10	0.38
Important achievements	0.17	0.13	0.86	0.14

Table 4.6 Output of Scientific and Technologic Funds per Million RMB in 2007

Item	Whole Country	Scientific Research Institutions	Colleges and Universities	Large and Medium-Sized Enterprises
Theses	0.65	0.43	6.25	0.04
Authorized patents	0.50	0.04	0.29	0.11
Important achievements	0.05	0.06	0.16	0.03

Table 4.7 Output of R&D Funds per Million RMB in 1999

Item	Whole Country	Scientific Research Institutions	Colleges and Universities	Large and Medium-Sized Enterprises
Theses	2.42	1.10	15.17	0.44
Authorized patents	0.43	0.08	0.15	1.08
Important achievements	0.40	0.29	1.28	0.40

Tables 4.9 through 4.12 show the outputs of theses and authorized patents of every 100 personnel engaging in scientific and technological activities and R&D. The outputs were higher in 2007 than in 1999. During 2007, the outputs of colleges and universities per 100 personnel engaging in scientific and technological activities and R&D personnel exceeded the outputs of scientific research institutions and large and medium-sized enterprises.

Table 4.8 Output of R&D Funds per Million RMB in 2007

Item	Whole Country	Scientific Research Institutions	Colleges and Universities	Large and Medium-Sized Enterprises
Theses	1.25	0.69	9.72	0.07
Authorized patents	0.95	0.06	0.45	0.21
Important achievements	0.09	0.09	0.24	0.06

Table 4.9 Outputs per 100 Persons Engaging in Scientific and Technological Activities in 1999

Item	Whole Country	Scientific Research Institutions	Colleges and Universities	Large and Medium-Sized Enterprises
Theses	4.74	4.24	25.19	0.47
Authorized patents	0.85	0.30	0.25	1.14
Important achievements	0.79	1.13	2.13	0.43

Table 4.10 Outputs per 100 Persons Engaging in Scientific and Technological Activities in 2007

Item	Whole Country	Scientific Research Institutions	Colleges and Universities	Large and Medium-Sized Enterprises
Theses	10.19	9.87	56.42	0.67
Authorized patents	7.74	0.88	2.60	1.98
Important achievements	0.75	1.31	1.40	0.55

If each thesis represents 1 point and each patent or achievement represents 10 points, we can calculate the integrated efficiency indices of the expenditures for science and technology by various research, educational, and commercial entities (Tables 4.13 and 4.14).

According to the integrated efficiency indices of scientific and technological expenditures for 2007, Chinese scientific research institutions, institutions of

Table 4.11 Outputs of Every 100 R&D Persons in 1999

Item	Whole Country	Scientific Research Institutions	Colleges and Universities	Large and Medium-Sized Enterprises
Theses	17.66	11.29	51.43	2.44
Authorized patents	3.17	0.80	0.51	5.93
Important achievements	2.94	3.01	4.34	2.23

Table 4.12 Output of Every 100 R&D Personnel in 2007

Item	Whole Country	Scientific Research Institutions	Colleges and Universities	Large and Medium-Sized Enterprises
Theses	26.68	15.81	120.39	1.72
Authorized patents	20.26	1.65	5.56	5.09
Important achievements	1.97	2.46	2.99	1.42

Table 4.13 Integrated Efficiency Indices of Scientific and Technological Expenditures in 1999

Index	Whole Country	Scientific Research Institutions	Institutions of Higher Education	Large and Medium-Sized Enterprises
Expenditures per million RMB for science and technology	4.63	2.09	19.83	5.36
Expenditures per million RMB for R&D	10.72	4.8	29.47	15.24
Distribution per 100 people engaged in scientific and technological activities	21.14	18.54	48.99	16.17
Distribution per 100 R&D people	78.76	49.39	99.93	84.04

Table 4.14   Integrated Efficiency Indices of Scientific and Technological Expenditures in 2007

Index	Whole Country	Scientific Research Institutions	Institutions of Higher Education	Large and Medium-Sized Enterprises
Expenditures per million RMB for science and technology	6.15	1.43	10.75	1.44
Expenditures per million RMB for R&D	11.65	2.19	16.62	2.77
Distribution per 100 people engaged in scientific and technological activities	95.09	31.77	96.42	25.97
Distribution per 100 R&D people	248.98	56.91	205.89	66.82

higher education, and large and medium-sized enterprises all showed increases in the efficiency of inputs and outputs of personnel and decreases in the input and output of funds compared with 1999. Analysis of inputs and outputs of technology in 2007 show the highest integrated efficiency among institutions of higher education, exceeding the index for the whole country and several times greater than the indices for scientific research institutions and large and medium-sized enterprises.

4.2.2 Conclusions and Recommendations

4.2.2.1 Conclusions and Problems

After analyzing the funds allocation structure for science and technology and expenditure and use efficiency, we came to the following conclusions:

1. As China's economic reformation continues, expenditures for science and technology showed gross and per capita increases. The funds allocation structure is undergoing revision. However, despite the need to use science and technology to build a moderately prosperous society, China's investment is inadequate, its investment structure is not sound, and basic science and technology operations are weak.
2. Business enterprises continually strengthen their economic positions and pursue technical innovation. Despite their scientific and technological outputs,

their capital use efficiency is still low and cooperation with professional organizations is minimal. Science and technology investments by banks and other financial institutions are declining. The current economic climate is not conducive for investments in science and technology.

3. The number of personnel employed in science and technology continues to grow and composition of the work force is more rational. Scientific and technological skills are improving as are per capita funding and employee efficiency.

4. National scientific and technological achievement shows obvious growth. The number of authorized patents of 2007 increased nearly 14-fold over 1999—a remarkable accomplishment. However, the use efficiency of funds is declining and needs improvement.

5. Scientific institutions always receive the most government funds for science and technology. Their reliance on government investment continues and funding increases. These institutions exhibit the best input–output efficiency. Allocations to educational institutes are still insufficient.

6. The system for allocating funds for science and technology is become more rational and the efficient use of funds is improving. However, few funds are available for construction and purchases of fixed assets.

4.2.2.2 Considerations and Recommendations

1. Build diversified and multichanneled science and technology investment systems. The government should play a leading role in investment via direct financing, tax benefits, and other fiscal steps. Scientific and technological activities should not be affected by market mechanisms.

2. Adjust and optimize the investment structure; increase the use efficiency of funds for science and technology. Strengthen public support for such activities. Build an effective investment management system. Improve the systems for distributing and monitoring invested funds. Devise a performance appraisal system for budgeting funds to companies and academic institutions and provide adequate supervision of expenditures for major projects.

3. Technological expenditure should be biased toward institutions of higher education. At present, their funding is far below funding of commercial enterprises and they use investment funds more efficiently. Cooperation among businesses, research institutions, and academic bodies should be strengthened. Commercial enterprises receive most scientific and technological investments. This should enhance their competitiveness and eventually allow better funding for other types of research entities.

4. Expenditure structures should be appropriately defined, reasonably planned, and constantly improved. A strict management system is critical to ensure

efficiency of research programs, define expenditure types and allocations, improve use efficiency, and eventually reduce investments. The budget for fundamental research should be increased.

5. According to a comparison of the input–output efficiency indices for science and technology for 1999 and 2007, the input–output efficiency indices for personnel of all types of institutions were observably higher although funding was lower. This shows that marginal benefit of per capita use of funds is decreasing. Future decisions for fund allocations should keep this fact in mind and per capita expenditures should be kept stable.

6. Build sharing mechanisms among organizations involved in scientific and technological activities to break the existing isolated, repetitious, and dispersive patterns and validate expenditures for science and technology.

4.3 Efficiency Evaluation of University Scientific and Technological Activities Based on DEA Model

In recent years, with the emphasis on technology innovation in China, the inputs for research activities increased sharply. In 2006, higher education research funding was 276.8 billion RMB—34.5 billion RMB more than in 2005 representing a 14.2% increase. Regional differences were greater. In 2006, 78 of the 129 universities applying to the Natural Science Foundation were in the eastern region (total funding = 1122.250 million RMB); 30 were in the central region (total funding = 371.798 million RMB); and 21 were in the western region (total funding = 198.691 million RMB). Outputs increased greatly. Patent applications in 2002 totaled 253,000; the number more than doubled in 2006 to 573,000 applications. Grants grew from 13.2 million RMB in 2002 to 26.8 million in 2006. Domestic scientific publications increased from 228,833 in 2002 to 404,858 in 2006; international publications increased from 77,395 in 2002 to 172,055 in 2006—more than double. China's economy is better developed in the eastern provinces—more institutions of higher learning, research funding, and total outputs. Investments and outputs in western regions are relatively small. Scientific and technological outputs lagged behind inputs and evaluating the efficiency of the use of funds for research is complicated. Data envelopment analysis (DEA) is suitable for that purpose.

4.3.1 Index Selection

An effective research funding assessment index system is required to establish research funding assessment criteria, compare the efficiencies of universities and research institutions funded by the scale, and ultimately improve the use of research funds. To reduce bias introduced by qualitative indices, we combined statistical methods to establish a quantitative analysis index system. A number of indices are identified by qualitative methods to develop a candidate index set. Evaluation indices are the selected from the candidate set by principal component and correlation analysis.

The first index to be selected must reflect the purpose and content of the evaluation index. The second index system must pay attention to refining and correlation issues. Finally, the system should consider the importance and availability of indices chosen.

There are m candidate input indices $x_1, x_2, \ldots x_m$, and n sample values indicated by $x_{ij}(i = 1,2,\ldots m, j = 1,2,\ldots n)$. $(x_{11}, x_{12}, \ldots x_{1n}), (x_{21}, x_{22}, \ldots x_{2n}), \ldots (x_{m1}, x_{m2}, \ldots x_{mn})$ are used to indicate a group of observations of m variables where $(x_{i1}, x_{i2}, \ldots x_{in})$ indicates a set of sample values $(i = 1,2,\ldots m)$ of the i-th variable that have n observations. The correlation matrix R can be calculated. $\lambda_1 \geq \lambda_2 \geq \ldots \geq \lambda_m \geq 0$ are m eigenvalues of R and $p_k = [p_{1k}, p_{2k}, \ldots p_{nk}]^T$ are the eigenvectors corresponding to λ_k. Via principal component analysis theory, we know that the k–th principal component is $z_k = \sum_{i=1}^{m} p_{ik} x_i$, and its variance contribution rate for m variables $x_{m1}, x_{m2}, \ldots x_{mn}$ is λ_k. If $\lambda_m \approx$, then $Rp_m \approx 0$ because $Rp_m = \lambda_m p_m$, that is, $XX^T p_m \approx 0$. Then $X^T p_m \approx 0$, namely $\sum_{i=1}^{m} p_{ik} x_i \approx 0$, which shows that $x_1, x_2, \ldots x_m$, are linear correlations. As the variance of x_k on λ_m is p_{km}^2, we can find one component that has the largest absolute value in p_m and remove the corresponding indices. Many trials indicate that the standard $\lambda_m \approx 0$ can be considered its value less than 0.01. When $\lambda_m \geq 0.01$, the analysis can stop.

The output indices should be highly correlated and concentrated. Suppose there are k candidate output indices denoted $y_1, y_2, \ldots y_k$. y_{ij} $(i = 1,2,\ldots k, j = 1,2,\ldots n)$ indicates n sample values. $(y_{11}, y_{12}, \ldots y_{1n})$, $(y_{21}, y_{22}, \ldots y_{2n})$, …, $(y_{k1}, y_{k2}, \ldots y_{kn})$ indicates a group of sample observations of k variables where $(y_{i1}, y_{i2}, \ldots y_{in})$ indicate a set of sample values of the i-th variable that has $n(i = 1,2,\ldots k)$ observations. A sample correlation matrix can be calculated. The output indices are selected by excluding indices whose correlation coefficients are small.

Indices selected should consider the purpose of the evaluation, requirements and characteristics of the object to be evaluated, and the expertise of the evaluators as determined by qualitative methods. The results of monitoring indices of scientific progress selected to analyze data from 2002 through 2006 from 31 provinces, autonomous regions, and municipalities are shown in Table 4.15.

4.3.2 Evaluation

DEA is suitable for evaluating relative effectiveness among different units. We used DEA to analyze the use of investment funds by colleges and universities for scientific research to determine relative efficiency.

Table 4.15   Input and Output Indices

	Index	Meaning
Input index	Government funds (X_1)	Funds allocated by government for scientific research
	Projects commissioned by enterprises (X_2)	Enterprise funding to schools and research institutions
	Full-time equivalent staff (X_3)	Engagement in research work or research applications, technology services accounting for 90% more of work hours during reference year (9 months or more excluding overtime and vacations)
Output index	Projects (Y_1)	Number of research projects for year
	Monographs (Y_2)	Number of monographs published during year
	Papers (Y_3)	Papers published in domestic and foreign journals during year
	Technology transfers (Y_4)	Scientific achievements transferred to another organization
	Awards (Y_5)	Government awards for accomplishments

$$
\begin{cases}
\operatorname{Min}\left[\theta - \varepsilon\left(e^T s^- {}_+^{\wedge} e^T s^+\right)\right] = v_D(\varepsilon) \\[2mm]
\text{s.t} \displaystyle\sum_{j=1}^{n} X_j(t)\lambda_j + s^- = \theta X_j(t) \\[2mm]
\displaystyle\sum_{j=1}^{n} Y_j(t+k)\lambda_j - s^+ = Y_0(t+k) \\[2mm]
\displaystyle\sum_{t=1}^{n} \lambda_t = 1 \\[2mm]
\lambda j \geq 0. j \in J, s^- \geq 0, s^+ \geq 0
\end{cases}
$$

$$
x_j(t) = (x_{1j}(t_1), x_{2j}(t_2), \ldots, x_{mj}(t_m))^T, Y_j(t+k)
$$

$$
= (y_{1j}(t_1+k_1), y_{2j}(t_2+k_2), \ldots, y_{sj}(t_s+k_s))^T
$$

where t_i, k_i and the last line of the equation above represent nonnegative integers; k is output relative to input lag. The lags are set according to the adjusted R^2 of multiple linear regressions. $k_1 = 1, k_2 = 2, k_3 = 1, k_4 = k_5 = 2$ represents the four lags. The θ denotes the efficiency of the j_0-th decision making unit (DMU), satisfying $0 \leq \theta \leq 1$. The economic implication of θ is the compression ratio of the input X in the event of Y as the output of X can be instead a linear combination of all j DMUs. Therefore, θ is also known as efficiency measurement value. When $\theta = 1$ a unit is efficient. When $\theta < 1$, a unit is inefficient; $1 - \theta$ is the extra input share of the j-th unit and the largest share that can be reduced. λ links every efficient point to form an effective frontier; nonzero S^+ and S^- allow the efficient surface to extend along the horizontal and vertical directions, forming an envelope face. In practice, nonzero slack variables are meaningful; θ indicates a radial optimal volume or distance required for a DMU to get close to the efficient frontier or envelopment surface.

4.3.3 Evaluation of Use of University Research Funds by Region

Using the Charnes, Cooper, and Rhodes (CCR) model and 2003–2007 *University Research Statistical Information* covering 3 regions of 31 provinces, autonomous regions, and municipalities, we determined the 5-year research funding DEA efficiency value (Table 4.16).

Table 4.16 shows effectiveness of decision making units of the provinces, municipalities, and autonomous regions. The western region includes Guizhou, Yunnan, Shaanxi, and Qinghai Provinces and the Tibet, Ningxia Hui, and Xinjiang Uygur Autonomous Regions. The eastern region includes Hebei, Zhejiang, and Fujian Provinces. The central region includes only Anhui and Henan Provinces.

Although Beijing is site of many key universities and research institutes, its DEA efficiency score is only 0.7002904. Beijing houses many government agencies, research institutes, and universities. It receives the most research investment but its regional research output is relatively small as a result of low input and output efficiency. In contrast, research investments in Ningxia, Qinghai, Hainan, Xinjiang, and Tibet are inadequate. Many of their outputs are created by universities in other provinces.

4.4 Evaluation of Regional Scientific and Technological Strength: Jiangsu Province

Scientific and technological strength is a comprehensive reflection of the economy, resources, research, technology, and other capabilities of a nation or region. Evaluating scientific and technological strength of a nation or region over a certain period can provide accurate measurements of scientific and technological capabilities related to inputs and outputs. Such evaluation requires development of a

Table 4.16 Efficiency Evaluation of Use of Scientific Research Funds by Provinces, Municipalities, and Autonomous Regions

DMU	DEA Value	Region	Scale	DMU	DEA Value	Region	Scale
Beijing	0.7003	East	Decreasing	Hubei Province	0.9810	Central	Decreasing
Tianjin	0.5268	East	Ascending	Hunan Province	0.7869	Central	Ascending
Hebei	1	East	Unchanged	Guangdong Province	0.8219	East	Ascending
Shanxi	0.5948	Central	Decreasing	Guangxi Zhuang Autonomous Region	0.6586	Central	Decreasing
Inner Mongolia Autonomous Region	0.7463	Center	Decreasing	Hainan Province	1	East	Unchanged
Liaoning Province	0.6046	East	Ascending	Chongqing	0.7564	West	Decreasing
Jilin Province	0.5066	Center	Decreasing	Sichuan Province	0.6870	West	Decreasing
Heilongjiang Province	0.5418	East	Decreasing	Guizhou Province	1	West	Unchanged
Shanghai	0.6042	East	Decreasing	Yunnan Province	1	West	Unchanged
Jiangsu Province	0.6834	East	Ascending	Tibet Autonomous Region	1	West	Unchanged
Zhejiang Province	1	East	Unchanged	Shanxi Province	1	West	Unchanged
Anhui Province	1	Central	Unchanged	Gansu Province	0.9205	West	Decreasing

(Continued)

Table 4.16 (Continued)   Efficiency Evaluation of Use of Scientific Research Funds by Provinces, Municipalities, and Autonomous Regions

DMU	DEA Value	Region	Scale	DMU	DEA Value	Region	Scale
Fujian Province	1	East	Unchanged	Qinghai Province	1	West	Unchanged
Jiangxi Province	0.7801	Central	Decreasing	Ningxia Hui Autonomous Region	1	West	Unchanged
Shandong Province	0.8023	East	Decreasing	Xinjiang Uygur Autonomous Region	1	West	Unchanged
Henan Province	1	Central	Decreasing				

complete index system and determination of the weight of each index. Evaluation of scientific and technology strength or weakness can serve as a basis for government decisions related to funding. This section discusses the comprehensive strength of science and technology of 31 Chinese provinces and 13 prefecture-level cities in Jiangsu Province determined by grey clustering.

Data are from the *China Statistical Yearbook* 2008, *China Science and Technology Statistical Yearbook, Statistical Yearbook of Jiangsu Province, Science and Technology Statistical Yearbook of Jiangsu Province*, and other sources.

4.4.1 Evaluation of Scientific and Technological Strength of China Provinces

4.4.1.1 Devising Evaluation Index System

Utilizing experts' suggestions and historical data, we established an index system to evaluate the scientific and technological strength of China's provinces. We analyzed science and technology inputs, activities, and outputs for 3 primary indices and 23 secondary indices as shown in Table 4.17.

4.4.1.2 Explanations of Indices

X1– Number of persons involved in science and technology activities, graduates of science, engineering, agriculture, and medical schools, persons in economic sectors committed to research, teaching, and production, engineering, agriculture, medicine, and related fields, and management professionals in research institutions, commercial enterprises, and education.

X2 – Number persons engaged in science and technology activities per 10,000 persons; X1 divided by total population.

X3 – Number of scientists and engineers who earned degrees or senior titles.

X4 – Number of scientists and engineers per 10,000 persons; X3 divided by total population.

X5 – Total expenditures for research and development activities (including basic and applied research and development) during the reporting period.

X6 – Per capita funds for science and technology activities; X5 divided by X1.

X7 – Proportion of R&D funds in GDP; X5 divided GDP.

X8 – Investment in research equipment by research institutions and colleges and universities.

X9 – Proportion of investment in fixed assets by research institution; X8 divided by (IFA + EFA + UFA). IFA = fixed asset construction spending by research institutions; EFA = fixed asset construction spending by medium-sized enterprises; UFA = fixed assets construction spending by institutions of higher learning.

X10 – Total retail sales of books.

Table 4.17   Index System for Assessing Scientific and Technological Strength

Primary Index	Secondary Index	Unit	Mark
Inputs	Persons involved in scientific and technological activities	Person	X1
	Persons involved in scientific and technological activities per 10,000 persons	Person	X2
	Scientists and engineers	Person	X3
	Scientists and engineers per 10,000 persons	Person	X4
	Total funds for scientific research	10,000 RMB	X5
	Per capita funds for scientific and technological activities	10,000 RMB per person	X6
	Proportion of R&D funds in GDP	Percent	X7
	Investments in research equipment	10,000 RMB	X8
	Proportion of investments in research equipment to total fixed assets of research institution	Percent	X9
	Book sales	Billion volumes	X10
	Number of computers per 100 families	Computer	X11
Activities	Numbers of research subjects	Subject	X12
	Numbers of state planning projects	Project	X13
	Numbers of university students per 10,000 residents	Person	X14
	Average years of education of workers	Years per person	X15
Outputs	Number of papers published	Paper	X16
	Number of granted patents	Patent	X17
	Technical market turnover	10,000 RMB	X18
	Index of economic benefits	Percent	X19

Table 4.17 (Continued)   Index System for Assessing Scientific and Technological Strength

Primary Index	Secondary Index	Unit	Mark
	Per capita GDP	RMB per person	X20
	Proportion of output value of new products to total products	Percent	X21
	Average staff wage	RMB	X22
	Per capita income of rural households	RMB	X23

X11 – Number of computers per 100 families.

X12 – Number of research subjects examined during the reporting period.

X13 – Number of state planning projects undertaken during the reporting period.

X14 – Number of university students per 10,000 persons; proportion of number of students to total population of province.

X15 – Average years of worker education; sum of education years divided by total population.

X16 – Number of scientific and technological papers published during the reporting period.

X17 – Number of patents granted during the reporting period.

X18 – Annual turnover of technology contracts.

X19 – Index of economic benefits; quality of economic operation for each province during the reporting period.

X20 – Per capita GDP; GDP divided by the population.

X21 – Proportion of the output value of new products to total industry outputs.

X22 – Average staff wage; total wage divided by number of staff members.

X23 – Per capita income of rural households; total peasant income divided by total number of peasants.

4.4.1.3 Concrete Values of Evaluation Indices for Scientific and Technological Strength of Provinces

For convenience, regions (provinces and cities) are numbered according to the *Statistical Yearbook*, as shown in Table 4.18.

The concrete values of 23 evaluation indices of the scientific and technological strength of 31 provinces are shown in Tables 4.19 through 4.21.

Table 4.18   Region Codes

1	Beijing	9	Shanghai	17	Hubei	25	Yunnan
2	Tianjin	10	Jiangsu	18	Hunan	26	Tibet
3	Hebei	11	Zhejiang	19	Guangdong	27	Shanxi
4	Shanxi	12	Anhui	20	Guangxi	28	Gansu
5	Inner Mongolia	13	Fujian	21	Hainan	29	Qinghai
6	Liaoning	14	Jiangxi	22	Chongqing	30	Ningxia
7	Jilin	15	Shandong	23	Sichuan	31	Xinjiang
8	Heilongjiang	16	Henan	24	Guizhou		

4.4.1.4 Grey Clustering Evaluation of Scientific and Technological Strength of China Provinces

Because indices have different meanings and their values differ widely, grey fixed weight clustering is applied. After coding the indices and assigning grey classifications, the whitenization weight function of X_j on the k-th class, $f_k^j(\bullet)$ ($j = 1,2,\ldots,31$; $k = 1,2,3$) can be defined where $k = 1,2,3$ corresponds to the strong, general, and weak grey classes, respectively.

$$f_1^1(140000,400000,-,-),\ f_1^2(100000,150000,-,200000),$$

$$f_1^3(-,-,70000,200000)$$

$$f_2^1(0.3,0.8,-,-),\ f_2^2(0.2,0.3,-,0.4),\ f_2^3(-,-,0.2,0.3)$$

$$f_3^1(80000,250000,-,-),\ f_3^2(50000,100000,-,150000),$$

$$f_3^3(-,-,40000,130000)$$

$$f_4^1(0.2,0.7,-,-),\ f_4^2(0.1,0.2,-,0.3),\ f_4^3(-,-,0.1,0.4)$$

$$f_5^1(1500000,500000,-,-),\ f_5^2(1000000,2000000,-,3000000),$$

$$f_5^3(-,-,1000000,2000000)$$

$$f_6^1(15,18,-,-),\ f_6^2(14,16,-,18),\ f_6^3(-,-,12,6)$$

$$f_7^1(2,3.8,-,-),\ f_7^2(1.3,2.4,-,3.5),\ f_7^3(-,-,1.3,2.5)$$

Table 4.19 Science and Technology Inputs

Region	X1	X2	X3	X4	X5	X6	X7	X8	X9	X10	X11
1	419741	2.48	335507	1.98	10888595	25.94	10.38	678089	73.70	0.74	85.93
2	123965	1.05	83883	0.71	3296804	26.59	5.19	30098	45.34	0.38	70.83
3	142628	0.20	98424	0.14	1904254	13.35	1.18	27078	89.61	1.48	55.27
4	133570	0.39	84853	0.25	1909986	14.30	2.75	28546	80.53	1.06	47.21
5	47997	0.20	34089	0.14	750417	15.63	0.97	7412	74.88	0.67	37.28
6	195465	0.45	141895	0.33	3530071	18.06	2.62	76840	78.93	1.36	51.2
7	97353	0.36	71150	0.26	1163858	11.96	1.81	13393	78.94	1.8	45.55
8	115777	0.30	82814	0.22	1507006	13.02	1.81	38468	64.97	0.54	37.91
9	224234	1.19	167899	0.89	5856520	26.12	4.28	238803	74.54	2.62	109
10	511670	0.67	323494	0.42	11735700	22.94	3.87	112118	47.73	4.21	68.17
11	413108	0.81	256324	0.50	7028269	17.01	3.27	44451	74.44	2.96	79.45
12	149049	0.24	98954	0.16	2514649	16.87	2.83	24770	66.70	2.79	50.73
13	130618	0.36	92346	0.26	2303021	17.63	2.13	17632	90.78	0.78	80.92
14	77340	0.18	51571	0.12	1131389	14.63	1.75	4657	13.18	1.51	50.66
15	363503	0.39	257752	0.27	7601016	20.91	2.45	47982	72.18	2.81	63.97

(Continued)

Table 4.19 (Continued) Science and Technology Inputs

Region	X1	X2	X3	X4	X5	X6	X7	X8	X9	X10	X11
16	206496	0.22	130142	0.14	2320122	12.69	1.42	79848	84.87	2.35	46.64
17	184072	0.32	131984	0.23	3102974	16.86	2.74	79556	75.62	2.32	51.89
18	147648	0.23	100294	0.16	2166358	14.67	1.94	20558	71.03	2.81	43.12
19	527477	0.55	385368	0.40	8504327	16.12	2.38	36684	56.58	2.8	83.23
20	67486	0.14	47869	0.10	860746	12.75	1.20	7750	62.98	2.57	67.62
21	10509	0.12	6582	0.08	146511	13.94	1.00	7721	71.69	0.68	47.52
22	87965	0.31	63095	0.22	1415627	16.09	2.78	6323	43.69	1.37	58.21
23	221582	0.27	141714	0.17	3615274	16.32	2.89	133065	70.70	1.95	49.14
24	39387	0.10	24898	0.07	509213	12.93	1.53	3613	43.08	1.04	43.37
25	63737	0.14	42906	0.09	842035	13.21	1.48	19820	49.71	1.68	40.19
26	3549	0.12	2430	0.08	54347	15.31	1.37	1995	78.88	0.13	29.95
27	147667	0.39	99085	0.26	2632252	17.83	3.84	160813	70.95	2.22	55.85
28	54031	0.21	37412	0.14	763444	14.13	2.40	27750	83.13	0.63	35.18
29	10879	0.20	7968	0.14	165558	15.22	1.72	570	76.61	0.08	35.31
30	14780	0.24	10397	0.17	178723	12.09	1.63	1133	74.59	0.15	38.56
31	34197	0.16	22131	0.10	539007	15.76	1.28	7290	79.92	0.84	41.32

Table 4.20   Science and Technology Activities

Region	X12	X13	X14	X15	Region	X12	X13	X14	X15
1	69195	285	6750	9.75	17	26017	303	2724	8.26
2	12984	103	4534	8.84	18	18534	185	1966	7.66
3	12045	165	1811	7.6	19	32143	669	1821	7.27
4	7203	167	1979	7.6	20	11914	107	1352	7.12
5	5014	144	1650	7.78	21	1867	41	1800	7.4
6	18830	290	2621	8.03	22	10242	145	2192	7.66
7	12954	203	2659	8.24	23	25644	294	1637	7.41
8	15373	281	2352	8.16	24	5839	81	969	6.9
9	35566	139	4371	8.99	25	9984	97	1174	7
10	32988	1742	2679	7.96	26	312	15	1279	7.36
11	34963	1858	2324	7.38	27	18646	198	2880	8.13
12	14551	312	1658	7.61	28	7402	247	1687	7.53
13	14789	234	1937	7.42	29	726	45	1033	7.17
14	11251	173	2062	7.42	30	2213	57	1610	7.3
15	22595	837	2071	7.71	31	3795	152	1414	7.27
16	12237	213	1648	7.38					

$$f_8^1(10000,100000,-,-),\ f_8^2(10000,30000,-,50000),\ f_8^3(-,-,3000,70000)$$

$$f_9^1(75,80,-,-),\ f_9^2(60,70,-,80),\ f_9^3(-,-,50,75)$$

$$f_{10}^1(1.4,2.7,-,-),\ f_{10}^2(1,2,-,3),\ f_{10}^3(-,-,0.4,1.8)$$

$$f_{11}^1(50,80,-,-),\ f_{11}^2(35,50,-,65),\ f_{11}^3(-,-,35,55)$$

$$f_{12}^1(12000,32000,-,-),\ f_{12}^2(9000,15000,-,21000),\ f_{12}^3(-,-,1000,18000)$$

$$f_{13}^1(200,600,-,-),\ f_{13}^2(100,250,-,400),\ f_{13}^3(-,-,100,200)$$

$$f_{14}^1(1800,4000,-,-),\ f_{14}^2(1600,2000,-,2400),\ f_{14}^3(-,-,1000,2000)$$

Table 4.21   Science and Technology Outputs

Region	X16	X17	X18	X19	X20	X21	X22	X23
1	41162	17747	10272173	5.15	57931	23.73	56328	10661.9
2	6009	6790	866122	7.44	48390	21.31	41748	7910.78
3	2876	5496	165906	6.04	19588	4.36	24756	4795.46
4	1423	2279	128425	7.2	16943	6.44	25828	4097.24
5	313	1328	94423	11.8	26128	3.71	26114	4656.18
6	10318	10665	997290	1.79	24945	7.46	27729	5576.48
7	5282	2984	196066	4.85	18126	18.17	23486	4932.74
8	7664	4574	412565	29.8	18763	5.93	23046	4855.59
9	19928	24468	3861695	3.51	95049	18.10	56565	11440.3
10	15659	44438	940246	6.62	33089	9.74	31667	7356.47
11	11016	52953	589189	4.78	36241	12.14	34146	9257.93
12	5784	4346	324865	6.1	11780	8.45	26363	4202.49
13	3131	7937	179690	6.92	30255	10.45	25702	6196.07
14	1183	2295	77641	4.79	12504	6.95	21000	4697.19
15	8216	26688	660126	6.77	29262	8.60	26404	5641.43
16	2766	9133	254425	7.91	15824	5.18	24816	4454.24
17	11994	8374	628971	7.89	15104	12.49	22739	4656.38
18	7427	6133	477024	6.61	13773	10.17	24870	4512.46
19	8363	62031	2016319	5.63	39951	11.37	33110	6399.79
20	887	2228	26996	4.39	13004	8.40	25660	3690.34
21	102	341	35602	5.96	13361	5.03	21864	4389.97
22	2532	4820	621884	5.71	16014	27.28	26985	4126.21
23	7682	13369	435313	6.31	11008	10.49	25038	4121.21
24	397	1728	20356	7.76	7264	5.83	24602	2796.93
25	1101	2021	50547	7.44	12184	5.07	24030	3102.6
26	7	93	0	6.13	13754	0.95	47280	3175.82

Table 4.21 (Continued)   Science and Technology Outputs

Region	X16	X17	X18	X19	X20	X21	X22	X23
27	10056	4392	438300	18.3	13284	6.87	25942	3136.46
28	2871	1047	297560	2.43	9827	6.53	24017	2723.79
29	86	228	77033	21.8	18940	4.46	30983	3061.24
30	50	606	8898	3.09	17540	5.89	30719	3681.42
31	341	1493	73963	26	17616	3.80	24687	3502.9

$$f_{15}^1(7.4,8.2,-,-),\ f_{15}^2(7.3,7.6,-,7.9),\ f_{15}^3(-,-,7.1,7.6)$$

$$f_{16}^1(6000,10000,-,-),\ f_{16}^2(1000,5000,-,9000),\ f_{16}^3(-,-,800,7000)$$

$$f_{17}^1(5000,13000,-,-),\ f_{17}^2(2000,5000,-,8000),\ f_{17}^3(-,-,600,6000)$$

$$f_{18}^1(200000,800000,-,-),\ f_{18}^2(100000,400000,-,700000),$$
$$f_{18}^3(-,-,30000,500000)$$

$$f_{19}^1(6,11,-,-),\ f_{19}^2(3,7,-,11),\ f_{19}^3(-,-,3,6.6)$$

$$f_{20}^1(15000,40000,-,-),\ f_{20}^2(10000,17000,-,24000),$$
$$f_{20}^3(-,-,10000,20000)$$

$$f_{21}^1(7,18,-,-),\ f_{21}^2(4,8,-,12),\ f_{21}^3(-,-,4,7)$$

$$f_{22}^1(26000,40000,-,-),\ f_{22}^2(23000,26000,-,29000),\ f_{22}^3(-,-,22000,27000)$$

$$f_{23}^1(4600,9000,-,-),\ f_{23}^2(3000,4600,-,6200),\ f_{23}^3(-,-,3000,4800)$$

After the index system established in Table 4.17 was scored by experts, the weight of each index could be obtained as shown in Table 4.22.

The fixed weight clustering coefficients matrix

$$\sum = (\sigma_i^k) = \begin{bmatrix} \sigma_1^1 & \sigma_1^2 & \sigma_1^3 \\ \dots\dots\dots\dots\dots \\ \sigma_{31}^1 & \sigma_{31}^2 & \sigma_{31}^3 \end{bmatrix}$$

Table 4.22 Weights of Indices

Index	X1	X2	X3	X4	X5	X6	X7	X8
Weight (η_i)	0.056016	0.05354	0.05973	0.05814	0.04412	0.05133	0.03922	0.04563
Index	X9	X10	X11	X12	X13	X14	X15	X16
Weight (η_i)	0.04424	0.04228	0.04129	0.04498	0.04972	0.03592	0.02171	0.03226
Index	X17	X18	X19	X20	X21	X22	X23	
Weight (η_i)	0.04481	0.03712	0.04187	0.04737	0.04172	0.03599	0.03097	

was could be obtained using the formula $\sigma_i^k = \Sigma_{j=1}^m f_j^k(x_{ij})\eta_j; i = 1,2,\cdots,31; k = 1,2,3$, Tables 4.20 through 4.22, and the results from the first two steps. The grey fixed weight clustering coefficients are shown in Table 4.23.

According to $\max_{1\leq k\leq 3}\{\sigma_i^k\} = \sigma_i^{k*}$, we can determine the grey category to which object i belongs as follows:

$$\max_{1\leq k\leq 3}\{\sigma_1^k\} = \sigma_1^1 = 0.8324,\ \max_{1\leq k\leq 3}\{\sigma_2^k\} = \sigma_2^1 = 0.5079,\ \max_{1\leq k\leq 3}\{\sigma_3^k\} = \sigma_3^3 = 0.4898$$

$$\max_{1\leq k\leq 3}\{\sigma_4^k\} = \sigma_4^2 = 0.3597,\ \max_{1\leq k\leq 3}\{\sigma_5^k\} = \sigma_5^3 = 0.6184,\ \max_{1\leq k\leq 3}\{\sigma_6^k\} = \sigma_6^1 = 0.4090$$

$$\max_{1\leq k\leq 3}\{\sigma_7^k\} = \sigma_7^2 = 0.3821,\ \max_{1\leq k\leq 3}\{\sigma_8^k\} = \sigma_8^2 = 0.5075,\ \max_{1\leq k\leq 3}\{\sigma_9^k\} = \sigma_9^1 = 0.8043$$

Table 4.23 Grey Fixed Weight Clustering Coefficients

I	σ_i^1	σ_i^2	σ_i^3	I	σ_i^1	σ_i^2	σ_i^3
1	0.8324	0.0842	0.0833	17	0.288	0.5707	0.1413
2	0.5079	0.2533	0.2388	18	0.0856	0.416	0.3384
3	0.0556	0.3067	0.4898	19	0.5681	0.2254	0.103
4	0.0635	0.3506	0.4232	20	0.0657	0.1803	0.754
5	0.0636	0.1038	0.6973	21	0	0.0139	0.8281
6	0.3314	0.5553	0.1133	22	0.0931	0.4806	0.4263
7	0.1177	0.4181	0.4642	23	0.2134	0.4341	0.2319
8	0.0996	0.4954	0.405	24	0.008	0.1012	0.8909
9	0.7635	0.148	0.0884	25	0.0046	0.1523	0.8431
10	0.7147	0.2331	0.0522	26	0.0703	0.095	0.8347
11	0.6784	0.2628	0.0588	27	0.2912	0.524	0.1847
12	0.1039	0.4615	0.3137	28	0.0501	0.1102	0.7104
13	0.2075	0.3551	0.2844	29	0.0689	0.1481	0.7829
14	0.0026	0.3046	0.6928	30	0.0121	0.0364	0.812
15	0.4722	0.4785	0.0492	31	0.0854	0.1414	0.7732
16	0.1416	0.3211	0.4084				

$$\max_{1 \le k \le 3}\left\{\sigma_{10}^k\right\} = \sigma_{10}^1 = 0.7505, \quad \max_{1 \le k \le 3}\left\{\sigma_{11}^k\right\} = \sigma_{11}^1 = 0.7269,$$

$$\max_{1 \le k \le 3}\left\{\sigma_{12}^k\right\} = \sigma_{12}^2 = 0.5585$$

$$\max_{1 \le k \le 3}\left\{\sigma_{13}^k\right\} = \sigma_{13}^2 = 0.3088, \quad \max_{1 \le k \le 3}\left\{\sigma_{14}^k\right\} = \sigma_{14}^3 = 0.5960,$$

$$\max_{1 \le k \le 3}\left\{\sigma_{15}^k\right\} = \sigma_{15}^1 = 0.5234$$

$$\max_{1 \le k \le 3}\left\{\sigma_{16}^k\right\} = \sigma_{16}^3 = 0.3339, \quad \max_{1 \le k \le 3}\left\{\sigma_{17}^k\right\} = \sigma_{17}^2 = 0.3996,$$

$$\max_{1 \le k \le 3}\left\{\sigma_{18}^k\right\} = \sigma_{18}^2 = 0.4731$$

$$\max_{1 \le k \le 3}\left\{\sigma_{19}^k\right\} = \sigma_{19}^1 = 0.6261, \quad \max_{1 \le k \le 3}\left\{\sigma_{20}^k\right\} = \sigma_{20}^3 = 0.7389,$$

$$\max_{1 \le k \le 3}\left\{\sigma_{21}^k\right\} = \sigma_{21}^3 = 0.7389$$

$$\max_{1 \le k \le 3}\left\{\sigma_{22}^k\right\} = \sigma_{22}^2 = 0.4762, \quad \max_{1 \le k \le 3}\left\{\sigma_{23}^k\right\} = \sigma_{23}^2 = 0.3492,$$

$$\max_{1 \le k \le 3}\left\{\sigma_{24}^k\right\} = \sigma_{24}^3 = 0.8025$$

$$\max_{1 \le k \le 3}\left\{\sigma_{25}^k\right\} = \sigma_{25}^3 = 0.7549, \quad \max_{1 \le k \le 3}\left\{\sigma_{26}^k\right\} = \sigma_{26}^3 = 0.7965,$$

$$\max_{1 \le k \le 3}\left\{\sigma_{27}^k\right\} = \sigma_{27}^2 = 0.4646$$

$$\max_{1 \le k \le 3}\left\{\sigma_{28}^k\right\} = \sigma_{28}^3 = 0.6262, \quad \max_{1 \le k \le 3}\left\{\sigma_{29}^k\right\} = \sigma_{29}^3 = 0.7567,$$

$$\max_{1 \le k \le 3}\left\{\sigma_{30}^k\right\} = \sigma_{30}^3 = 0.7506$$

$$\max_{1 \le k \le 3}\left\{\sigma_{31}^k\right\} = \sigma_{31}^3 = 0.6728$$

Beijing, Tianjin, Liaoning, Shanghai, Jiangsu, Zhejiang, Shandong, and Guangdong rated best in scientific and technological strength. Shanxi, Jilin, Heilongjiang, Anhui, Fujian, Hubei, Hunan, Chongqing, Sichuan, and Shanxi belong to the general class. Hebei, Inner Mongolia, Jiangxi, Henan, Guangxi, Hainan, Guizhou, Yunnan, Xizang, Gansu, Qinghai, Ningxia, and Xinjiang fall into the weak class.

4.4.2 Evaluation of Scientific and Technological Strengths of Prefecture-Level Cities in Jiangsu Province

4.4.2.1 Designing Evaluation Index System

Using experts' suggestions and historical data, we established the evaluation index system to assess the scientific and technological strength for prefecture-level cities. We examined science and technology inputs, activities, and outputs for the 3 primary indices and the 23 secondary indices as shown in Table 4.24.

4.4.2.2 Explanations of Indices

$X1$ – Number of persons involved in science and technology activities including graduates of science, engineering, agriculture, and medical schools, persons in economic sectors committed to research, teaching, and production, engineering, agriculture, medicine, and related fields, and management professionals in research institutions, commercial enterprises, and education.

$X2$ – Proportion of persons engaged in technology activities to total employees; $X1$ divided by the total number of employees.

$X3$ – Proportion of persons engaging in R&D to total workers in science and technology activities; number of R&D workers divided by $X1$.

$X4$ – Proportion of the persons engaging in R&D activities to total staff; number or R&D workers divided by total staff.

$X5$ – Proportion of government technology grants to total financial expenditures; number of grants divided by total expenditures.

$X6$ – Proportion of corporate R&D expenditures from sales revenues; R&D expenditures divided by sales revenues.

$X7$ – Proportion of R&D expenditure to GDP; R&D expenditures divided by GDP.

$X8$ – Number of granted patents per 100,000 population; total number of granted patents granted during the reporting period divided by total population.

$X9$ – National average education years; sum of education years divided by total population.

$X10$ – Number of students in secondary and higher schools per 10,000 persons; number of such students divided by total population.

$X11$ – Per capita GDP; GDP divided by total population.

$X12$ – GDP growth rate; GDP for this year and last year divided by GDP of last year.

$X13$ – Sales revenues of high technology industries; total sales revenue for reporting period.

$X14$ – Proportion of output value of high technology industries to total industry outputs; outputs of high technology industries divided by total industrial output.

Table 4.24   Evaluation Index System for Assessing Scientific and Technological Strength

Primary Index	Secondary Index	Unit	Mark
Inputs	Number of persons engaged in scientific and technological activities	10,000 persons	X1
	Proportion of persons engaged in scientific and technological activities to total staff	Percent	X2
	Proportion of R&D workers to workers engaged in scientific and technological activities	Percent	X3
	Proportion of R&D workers to total staff	Percent	X4
	Proportion funds appropriated for science and technology by the government	Percent	X5
	Proportion of corporate R&D expenditure to sales revenue	Percent	X6
	Proportion of R&D expenditures to GDP	Percent	X7
Activities	Number of granted patents per 100,000 persons	Patents per 10,000 persons	X8
	National average years of education	Years per person	X9
	Number of students in secondary and higher schools per 10,000 persons	Students per 10,000 persons	X10
Outputs	Per capita GDP	RMB	X11
	GDP growth rate	Percent	X12
	Sales revenues of high technology industries	Billion RMB	X13
	Proportion of outputs of high technology industries to total industry	Percent	X14

(Continued)

Table 4.24 (Continued)   Evaluation Index System for Assessing Scientific and Technological Strength

Primary Index	Secondary Index	Unit	Mark
	Contribution of high technology industries to increased industrial output	Percent	X15
	Labor productivity	RMB per person	X16
	Industrial energy consumption per 10,000 RMB added value	10,000 RMB per ton of standard coal	X17
	Total power of agricultural machinery per 1,000 acres of effective irrigation area	1,000 billion hours per 1,000 acres	X18
	Average wages of staff and workers	RMB	X19
	Per capita net income of rural residents	RMB	X20

X15 – Contribution rate of high technology industries to increased industrial output; increase of outputs of high technology industries divided by total increase of industrial output.

X16 – Labor productivity; industrial output divided by average total staff.

X17 – Energy consumption per 10,000 RMB industrial added value; added value RMB divided by total energy consumption.

X18 – Total power of agricultural machinery per thousand acres of effective irrigation area; total of all kinds of power consumed by machinery used in agriculture, forestry, animal husbandry, and fisheries. Effective irrigation areas are level, cultivated, have water supplies or irrigation systems.

X19 – Average wages of staff and workers; total wages divided by number of staff and workers.

X20 – Per capita net income of rural residents; total incomes of rural residents divided by population.

4.4.2.3 Concrete Values of Evaluation Indices

For convenience, prefecture-level cities in Jiangsu Province are numbered according to the *Statistical Yearbook*, as shown in Table 4.25.

The concrete values of 20 evaluation indices of scientific and technological strength for 13 prefecture-level cities in Jiangsu Province are shown in Tables 4.26 through 4.28.

Table 4.25 Codes of Prefecture-Level Cities in Jiangsu Province

1	Nanjing	6	Nantong	10	Yangzhou
2	Wuxi	7	Lianyungang	11	Zhenjiang
3	Xuzhou	8	Huaian	12	Taizhou
4	Changzhou	9	Yancheng	13	Suqian
5	Suzhou				

Table 4.26 Scientific and Technological Inputs of Prefecture-Level Cities in Jiangsu Province

Region	X1	X2	X3	X4	X5	X6	X7
1	71.23	2.34	51.49	4.21	2.34	0.88	2.45
2	47.21	1.48	61.32	2.35	2.39	0.98	2.29
3	24.83	0.43	58.11	1.86	1.34	0.79	1.32
4	27.54	2.13	45.91	3.28	2.95	0.88	2.27
5	68.16	2.1	43.02	1.46	3.66	0.68	1.94
6	31.53	0.87	42.34	2.29	2.59	0.7	1.56
7	18.53	0.3	54.61	2.02	2.09	0.85	1.13
8	6.83	0.19	46.69	1.23	2	0.6	0.85
9	20.48	0.43	47.34	1.37	1.53	0.43	0.74
10	18.69	0.92	50	2.13	2.42	0.77	1.76
11	15.04	1.46	48.83	2.29	2.56	0.88	1.77
12	23.59	0.66	49.24	2.06	2.15	0.8	1.62
13	12.72	0.1	39.86	0.67	1.4	0.38	0.34

4.4.2.4 Grey Clustering Evaluation

To accommodate different significances of clustering indices and disparities of data, we used the grey fixed weight clustering method. After coding the indices and grey classifications, the whitenization weight function of X_j on the k-th class $f_j^k(\bullet)(j = 1,2,\ldots,31; k = 1,2,3)$ can be defined where $k = 1,2,3$ correspond to the strong, general, and weak grey classes, respectively.

$$f_1^1(18,47,-,-), \ f_1^2(10,20,-,30), \ f_1^3(-,-,12,25)$$

Table 4.27 Science and Technology Activities of Prefecture-Level Cities in Jiangsu Province

Region	X8	X9	X10	Region	X8	X9	X10
1	63.08	9.520828	1053.9	8	9.23	8.027754	224.12
2	78.33	8.342688	305.37	9	16.26	8.497468	107.15
3	20.04	9.027228	209.5	10	44.8	8.246995	263.24
4	57.54	8.421277	321.3	11	84.88	8.502405	367.35
5	200.19	8.359705	273.49	12	32.83	7.890024	116.91
6	57.39	8.058857	185.65	13	3.78	7.880797	226.69
7	7.68	7.754823	181.79				

$$f_2^1(0.5,2,-,-),\ f_2^2(0.2,0.9,-,1.6),\ f_2^3(-,-,0.2,0.4)$$

$$f_3^1(45,54,-,-),\ f_3^2(43,48,-,53),\ f_3^3(-,-,42,49)$$

$$f_4^1(2,3,-,-),\ f_4^2(1.3,2.3,-,3.3),\ f_4^3(-,-,1.2,2)$$

$$f_5^1(2,2.5,-,-),\ f_5^2(1.5,2.3,-,3.1),\ f_5^3(-,-,1.4,2.4)$$

$$f_6^1(0.7,0.8,-,-),\ f_6^2(0.7,0.8,-,0.9),\ f_6^3(-,-,0.4,0.7)$$

$$f_7^1(1.5,2,-,-),\ f_7^2(1.1,1.7,-,2.3),\ f_7^3(-,-,0.7,2.5)$$

$$f_8^1(50,80,-,-),\ f_8^2(20,50,-,80),\ f_8^3(-,-,10,60)$$

$$f_9^1(8.1,9,-,-),\ f_9^2(8,8.3,-,8.6),\ f_9^3(-,-,7.8,8.4)$$

$$f_{10}^1(200,300,-,-),\ f_{10}^2(180,220,-,260),\ f_{10}^3(-,-,180,260)$$

$$f_{11}^1(25000,50000,-,-),\ f_{11}^2(20000,35000,-,50000),\ f_{11}^3(-,-,17000,50000)$$

$$f_{12}^1(16,20,-,-),\ f_{12}^2(14,17,-,20),\ f_{12}^3(-,-,15,19)$$

$$f_{13}^1(750,2500,-,-),\ f_{13}^2(350,800,-,1250),\ f_{13}^3(-,-,300,800)$$

Table 4.28 Science and Technology Outputs of Prefecture-Level Cities in Jiangsu Province

Region	X11	X12	X13	X14	X15	X16	X17	X18	X19	X20
1	50327	15.27	2510.10	41.3	40.88	99245.47	2.11	1.10	39876	8951
2	73053	14.55	2588.69	26.16	22.39	126264.2	1.08	0.86	38843	11280
3	23069	19.52	285.11	10.27	17.53	41299.45	2.45	1.06	26824	6240
4	50283	17.14	1484.10	29.26	33.91	75737.87	1.11	0.96	34834	10171
5	74676	17.55	5904.16	34.8	45.24	129478.5	1.07	0.81	36090	11785
6	35040	14.12	1359.77	27.14	38.31	52981.53	0.8	0.67	30856	7811
7	16808	21.92	186.18	21.25	25.07	27167.69	1.45	1.02	26596	5454
8	18921	19.68	131.75	11.21	19.9	29256.01	2.1	1.13	23993	5657
9	21238	16.92	370.37	15.97	24.97	47700.46	0.97	0.78	22380	6867
10	35232	20	814.19	24.74	31.49	57545.35	1	0.76	27323	7450
11	46473	16.09	794.60	30.62	39.52	85140.58	1.43	1.06	30958	8703
12	30256	15.97	791.74	28.92	28.15	49599.77	0.98	0.75	25737	7338
13	13709	20.86	21.99	3.9	5.32	21139.15	1.03	1.45	19988	5406

$$f_{14}^1(25,30,-,-),\ f_{14}^2(22,26,-,30),\ f_{14}^3(-,-,15,28)$$

$$f_{15}^1(20,35,-,-),\ f_{15}^2(22,30,-,38),\ f_{15}^3(-,-,19,34)$$

$$f_{16}^1(30000,90000,-,-),\ f_{16}^2(30000,55000,-,70000),\ f_{16}^3(-,-,25000,80000)$$

$$f_{17}^1(1,2,-,-),\ f_{17}^2(1,1.5,-,2),\ f_{17}^3(-,-,1.4,2)$$

$$f_{18}^1(0.8,1.1,-,-),\ f_{18}^2(0.8,1,-,1.2),\ f_{18}^3(-,-,0.7,1)$$

$$f_{19}^1(27000,36000,-,-),\ f_{19}^2(25000,30000,-,35000),\ f_{19}^3(-,-,22000,34000)$$

$$f_{20}^1(7000,10000,-,-),\ f_{20}^2(6000,7400,-,8800),\ f_{20}^3(-,-,5500,8000)$$

After the index system established in Table 4.17 was scored by experts, the weight of each index could be determined as shown in Table 4.29.

The fixed weight clustering coefficients matrix

$$\sum = (\sigma_i^k) = \begin{bmatrix} \sigma_1^1 & \sigma_1^2 & \sigma_1^3 \\ \cdots\cdots\cdots\cdots\cdots\cdots \\ \sigma_{31}^1 & \sigma_{31}^2 & \sigma_{31}^3 \end{bmatrix}$$

Was obtained from the formula $\sigma_i^k = \sum_{j=1}^m f_j^k(x_{ij})\cdot\eta_i$; $i=1,2,\cdots,31$; $k=1,2,3$, using the data from Tables 4.26 through 4.28, the whitenization weight function, and the weights of the indices. The grey fixed weight clustering coefficients are shown in Table 4.30.

Table 4.29　Weights of Indices

Index	X1	X2	X3	X4	X5	X6	X7
Weight	0.057	0.057	0.058	0.054	0.049	0.049	0.05
Index	X8	X9	X10	X11	X12	X13	X14
Weight	0.052	0.045	0.051	0.053	0.045	0.049	0.047
Index	X15	X16	X17	X18	X19	X20	
Weight	0.047	0.047	0.048	0.049	0.045	0.048	

Table 4.30 Grey Fixed Weight Clustering Coefficients

I	σ_i^1	σ_i^2	σ_i^3	i	σ_i^1	σ_i^2	σ_i^3
1	0.877	0.1464	0.0463	8	0.1616	0.1454	0.6856
2	0.6948	0.2252	0.1693	9	0.0796	0.2752	0.6679
3	0.3042	0.277	0.4458	10	0.3325	0.6618	0.2615
4	0.6879	0.2625	0.1142	11	0.6206	0.4503	0.1613
5	0.6677	0.1205	0.2034	12	0.2188	0.5651	0.4041
6	0.2799	0.4355	0.3646	13	0.1091	0.0609	0.867
7	0.2368	0.2801	0.5763				

According to $\max\limits_{1\le k\le 3}\{\sigma_i^k\} = \sigma_i^{k^*}$, we can determine the grey category to which object i belongs as follows.

$$\max_{1\le k\le 3}\left\{\sigma_1^k\right\} = \sigma_1^1 = 0.8770,\ \max_{1\le k\le 3}\left\{\sigma_2^k\right\} = \sigma_2^1 = 0.6948,\ \max_{1\le k\le 3}\left\{\sigma_3^k\right\} = \sigma_3^3 = 0.4458$$

$$\max_{1\le k\le 3}\left\{\sigma_4^k\right\} = \sigma_4^1 = 0.6879,\ \max_{1\le k\le 3}\left\{\sigma_5^k\right\} = \sigma_5^1 = 0.6677,\ \max_{1\le k\le 3}\left\{\sigma_6^k\right\} = \sigma_6^2 = 0.4355$$

$$\max_{1\le k\le 3}\left\{\sigma_7^k\right\} = \sigma_7^3 = 0.5763,\ \max_{1\le k\le 3}\left\{\sigma_8^k\right\} = \sigma_8^3 = 0.6856,\ \max_{1\le k\le 3}\left\{\sigma_9^k\right\} = \sigma_9^3 = 0.6679$$

$$\max_{1\le k\le 3}\left\{\sigma_{10}^k\right\} = \sigma_{10}^2 = 0.6618,\ \max_{1\le k\le 3}\left\{\sigma_{11}^k\right\} = \sigma_{11}^1 = 0.6206,\ \max_{1\le k\le 3}\left\{\sigma_{12}^k\right\} = \sigma_{12}^2 = 0.5651$$

$$\max_{1\le k\le 3}\left\{\sigma_{13}^k\right\} = \sigma_{13}^3 = 0.8670$$

In summary, the cities of Nanjing, Wuxi, Changzhou, Suzhou, and Zhenjiang belong to the strong class. Nantong, Yangzhou, and Taizhou belong to the general class. Xuzhou, Lianyungang, Huaian, Yancheng, and Suqian are in the weak class based on the results of grey fixed weight clustering of scientific and technological strength.

Chapter 5

Evaluation of Energy Saving in China

5.1 Introduction

During the 21st century, China's high energy-consuming industries such as iron, steel, and cement expanded rapidly, accelerating the development of the heavy machinery and chemical industries that accounted for a corresponding increase in growth of the national economy. China's economy expanded about 10% with a concomitant energy consumption increase. The growth rate of energy consumption exceeded the growth rate of the gross domestic product (GDP). The rapid growth of energy consumption has constrained sustainable development in China and even represents a threat to the nation's economic security. The energy situation in China is serious and a consensus at all levels of society indicates that energy-saving measures that will aid the nation to sustain economic development are critical.

The Chinese government has pursued several measures to decrease energy intensity. The National Development and Reform Commission announced its "Energy Development Eleventh Five-Year Plan" in April 2007 and issued a "Medium and Long Term Development Plan for Renewable Energy" in September 2007citing energy-saving and nonfossil energy development goals. The revised "Energy Conservation Law" was approved by the National People's Congress in October 2007. It sets specific energy-saving regulations and responsibilities and functions as the main basis of China's energy-saving system. In June 2007, the State Council-issued "Comprehensive Energy-Saving Program" was drawn up by the National Development and Reform Commission and other departments. The program proposes specific measures to save energy and reduce emissions. In November 2007, the State Council issued "Energy-Saving Statistics, Monitoring and Assessment Programs" and approaches to implement them. The Ministry of Finance issued "Interim Regulations for Special Funds for Renewable Energy Development" to support the development of renewable energy.

An energy evaluation provides an objective basis for a government to develop an energy policy and evaluate the effects of energy management scientifically. This chapter evaluates technological progress, industrial structure adjustments, the efficiency of nonfossil energy sources, and energy policy.

5.2 Energy-Saving Effects of Technological Progress

The ongoing industrialization and urbanization of China increased secondary industrial development. The energy-saving effect of the change of the industrial structure is relatively small. Industry consumes about 50% of China's energy. Therefore, industry represents a logical sector to investigate to determine the energy-saving effects of technological progress using the extended Cobb–Douglas production function.

5.2.1 Extended Cobb–Douglas Production Function

The histories of the economic growth in developed countries show that capital, labor, energy, and technological progress are the basic elements. The economic growth model focuses on five variables: output (Y), capital (K), labor (L), energy (E), and technological progress (T). Capital, labor, energy, and technological progress are combined to produce output. The production function takes the form

$$Y(t) = f(K(t), L(t), E(t), T(t)) \tag{5.1}$$

Assume that technological progress is exogenous and has a constant growth rate c; then technological progress grows exponentially

$$T(t) = Ae^{ct} \tag{5.2}$$

The Cobb–Douglas production function is easy to analyze and yields a good approximation of actual production (Romer, 2001). The production function is:

$$Y(t) = Ae^{ct} K(t)^{\alpha} L(t)^{\beta} E(t)^{\gamma} \tag{5.3}$$

where:
α = elasticity of output of capital
β = elasticity of output of labor

γ = elasticity of output of energy

$0 < \alpha, \beta, \gamma < 1$

Assume that the production function has constant scales for these outputs:

$$\alpha + \beta + \gamma = 1 \qquad (5.4)$$

From (5.4), we get

$$\left(\frac{E}{Y}\right)^{\gamma} \times e^{ct} \times A = \frac{Y^{1-\gamma}}{K^{\alpha} L^{\beta}} \qquad (5.5)$$

Since $\alpha + \beta + \gamma = 1$, (5.5) is changed to

$$\left(\frac{E}{Y}\right)^{\gamma} \times e^{ct} \times A = \left(\frac{Y}{K}\right)^{\alpha} \left(\frac{Y}{L}\right)^{\beta} \qquad (5.6)$$

where:

$\chi = E/Y$ indicates energy consumption per output or energy intensity

$y_k = Y/K$ Indicates output per capital unit

$y_l = Y/L$ indicates output per labor unit

Thus, (5.6) can be rewritten as

$$\chi^{\gamma} \times e^{ct} \times A = y_k^{\alpha} y_l^{\beta} \qquad (5.7)$$

We calculate the natural log of the two sides of (5.7) to yield:

$$\gamma \ln \chi(t) + ct + \ln A = \alpha \ln y_k(t) + \beta \ln y_l(t) \qquad (5.8)$$

We then calculate the derivative of the two sides of (5.8), yielding:

$$\gamma \frac{\dot{\chi}(t)}{\chi(t)} = \alpha \frac{\dot{y}_k(t)}{y_k(t)} + \beta \frac{\dot{y}_l(t)}{y_l(t)} - c \qquad (5.9)$$

where:

$\dot{\chi}(t)/\chi(t)$ = growth rate of $\chi(t)$ (energy intensity)

$\dot{y}_k(t)/y_k(t)$ = growth rate of $yk(t)$ (output per capital unit)

$\dot{y}_l(t)/y_l(t)$ = growth rate of $yl(t)$ (output per labor unit)

C = growth rate of technological progress

Equation (5.9) shows that the growth rate of energy intensity was based on the growth rates of output per capital unit, output per labor unit, elasticity of outputs, and technological progress. The larger the growth rate of capital and labor outputs, the larger the energy intensity. The larger the growth rate of technological progress, the smaller the energy intensity. When technological progress exceeds output, energy intensity will decrease.

5.2.2 Data

Industry is an important sector of the Chinese economy. The value added of industry exceeds 40% of the GDP. Industry accounts for more than 70% of total energy consumption in China and thus is an example that should be studied. Table 5.1 shows annual average balances of net value of fixed assets, averages of employed persons, energy consumption, and industry value added for 1995 through 2006. Table data reflect current values and must be adjusted to reflect constant values. The GDP deflator should be calculated before the value added at constant price is calculated.

Table 5.1   Annual Average Balances of Net Value of Fixed Assets, Employed Persons, Energy Consumption, and Value Added by Industry, 1995–2006

Year	Annual Average Balance of Net Value of Fixed Assets (10 Billion Yuan)	Annual Average Employed Persons (100,000 Persons)	Energy Consumption (Million tce)	Value Added of Industry (10 Billion Yuan)
1995	274.23	661.00	961.91	154.46
1996	378.63	645.00	1003.22	180.26
1997	397.79	621.50	1000.80	198.35
1998	454.21	619.5.8	944.09	194.22
1999	476.43	580.51	907.97	215.65
2000	519.10	555.94	896.34	253.95
2001	564.87	544.14	923.47	283.29
2002	607.98	552.07	1021.81	329.95
2003	714.88	574.86	1196.27	419.90
2004	868.85	609.86	1432.44	548.05
2005	894.60	689.60	1594.92	721.87
2006	1058.05	735.84	1751.37	910.76

Source: China Statistical Yearbook, 1996–2007.

$$\text{Indices of GDP} = \frac{\text{GDP at current price}/\text{GDP deflator}}{\text{GDP of preceding year}} \quad (5.10)$$

So

$$\text{GDP deflator} = \frac{\text{GDP at current price}/\text{indices of GDP}}{\text{GDP of preceding year}} \quad (5.11)$$

GDP at current price, indices of GDP, and GDP deflator calculated by Equation (5.12) are shown in Table 5.2. Annual average balance of net value of fixed assets at a constant price is calculated according to the investment of fixed assets price index as shown in Table 5.2.

$$\text{Value added at constant price} = \frac{\text{Value added at current price}}{\text{GDP deflator}} \quad (5.12)$$

Table 5.2   GDP Deflator and Investment of Fixed Assets Price Index

Year	GDP (10 Billion Yuan)	GDP Index	GDP Deflator	Investment of Fixed Assets Price Index
1995	60.79	–	–	–
1996	71.18	110	1.06	104
1997	78.97	109.3	1.02	101.7
1998	84.40	107.8	0.99	99.8
1999	89.68	107.6	0.99	99.6
2000	99.21	108.4	1.02	101.1
2001	109.66	108.3	1.02	100.4
2002	120.33	109.1	1.01	100.2
2003	135.82	110	1.03	102.2
2004	159.88	110.1	1.07	105.6
2005	183.87	110.4	1.04	101.6
2006	210.87	111.1	1.03	101.5

Source: China Statistical Yearbook, 2007.

Table 5.3 Annual Average Balance of Net Value of Fixed Assets and Value Added of Industry

Year	Annual Average Balance of Net Value of Fixed Assets (10 Billion Yuan)	Value Added of Industry (10 Billion Yuan)
1995	274.23	154.46
1996	364.07	169.38
1997	376.09	183.59
1998	430.29	181.38
1999	453.17	203.99
2000	488.38	235.43
2001	529.32	257.35
2002	568.58	297.95
2003	654.16	369.61
2004	752.89	451.16
2005	763.00	570.62
2006	889.07	697.34

Annual average balance of net value of fixed assets at constant price

$$= \frac{\text{annual average balance of net value of fixed assets at current price}}{\text{investment of fixed assets price index}} \quad (5.13)$$

The values at constant prices calculated via Equations (5.12) and (5.13) are shown in Table 5.3.

Due to economic shock waves, the data indicate that development was too fast or too slow; these changes do not reflect the true development trends of the economic system. Such interference should be eliminated if the conclusions revealed by models are to be unbelievable.

The Asian financial crisis occurred in 1997 and affected investment, energy consumption, and economic growth as shown in Figure 5.1. Annual average balance of net value of fixed assets and value added grew very slowly from 1997 to 2001. Energy consumption and annual average employment decreased continuously from 1997 to 2000.

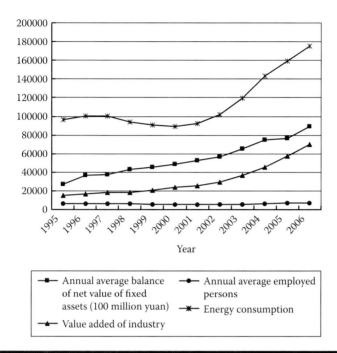

X-axis: Year (1995–2006)

Legend:
- Annual average balance of net value of fixed assets (100 million yuan)
- Value added of industry
- Annual average employed persons
- Energy consumption

Figure 5.1   Impact of Asian financial crisis on Chinese industry.

The average weakened buffer operator (AWBO) technique shown in Definition 5.1 has been applied widely in modeling and predictions for systems subject to shock waves such as the 1997 crisis and realistically analyze the data collected.

Definition 5.1

AWBO

Assume the original data sequence is

$$X = (x(1), x(2), \cdots, x(n))$$

Setting

$$XD = (x(1)d, x(2)d, \cdots, x(n)d)$$

Here $k = 1, 2, \ldots, n$

$$x(k)d = \frac{1}{n-k+1}[x(k) + x(k+1) + \cdots + x(n)],$$

Table 5.4 Annual Average Balance of Net Value of Fixed Assets and Value Added of Industry Based on ABWO

K'^a	Y'^b	E'^c
545.27	314.35	1202.00
569.91	328.89	1220.07
590.49	344.84	1242.00
614.32	362.76	1275.11
637.32	385.43	1321.00
663.63	411.35	1381.66
692.84	440.67	1458.03
725.54	477.34	1545.27
764.78	522.18	1632.53
801.65	573.04	1699.22
826.04	633.98	1751.37
889.07	697.34	1751.37

[a] K' indicates annual average balance of net value of fixed assets subjected to ABWO.
[b] Y' indicates value added subjected to ABWO.
[c] E' indicates energy consumption subjected to ABWO.

Then D is the AWBO. The annual average balance of net value of fixed assets, value added, and energy consumption subjected to AWBO, $K' = KD$, $Y' = YD$ and $E' = ED$ produced the results shown in Table 5.4. The data analyzed via BWO grow smoothly.

5.2.3 Empirical Research

The energy-saving effects of technological progress by Chinese industry are calculated as shown below. We first calculate natural log of the sides of (5.3) yielding

$$\ln Y = \ln A + ct + \alpha \ln K + \beta \ln L + \gamma \ln E \qquad (5.14)$$

According to (5.4),

$$\beta = 1 - \alpha - \gamma \qquad (5.15)$$

Table 5.5   Variables for Parameter Estimation

	t	*lnKL*[a]	*lnEL*[b]	*lnYL*[c]
1995	1	2.11	2.90	1.56
1996	2	2.18	2.94	1.63
1997	3	2.25	2.99	1.71
1998	4	2.29	3.02	1.77
1999	5	2.40	3.12	1.89
2000	6	2.48	3.21	2.00
2001	7	2.54	3.29	2.09
2002	8	2.58	3.33	2.16
2003	9	2.59	3.35	2.21
2004	10	2.58	3.33	2.24
2005	11	2.48	3.23	2.22
2006	12	2.49	3.17	2.25

[a] $\ln KL = \ln K - \ln L$.
[b] $\ln EL = \ln E - \ln L$.
[c] $\ln YL = \ln Y - \ln L$.

And from (5.14) and (5.15), we get

$$\ln Y - \ln L = \ln A + ct + \alpha(\ln K - \ln L) + \gamma(\ln E - \ln L) \qquad (5.16)$$

Equation (5.16) is a linear regression model. When the parameters are estimated, *Y, E,* and *K* will be replaced by *Y′, E′,* and *K′*. The variables required for parameter estimation are shown in Table 5.5.

The parameters are estimated by SPSS, and the results of coefficients and related tests are shown in Tables 5.6 and 5.7. The linear regression equation is:

$$\ln Y - \ln L = -0.24 + 0.044t + 0.269(\ln K - \ln L) + 0.407(\ln E - \ln L) \qquad (5.17)$$

The adjusted R^2 of 0.999 and significant F change of 0.000 show that the regression model is good and independent variables explain dependent variable very well. A t-test shows that $\ln K - \ln L$, and $\ln E - \ln L$ have significant effects on $\ln Y - \ln L$.

Table 5.6 R^2 **Test**

R	R^2	Adjusted R^2	R^2 Change	F Change	Significant F Change
1.000a	0.999	0.999	0.999	4698.636	0.000

Table 5.7 **Coefficients and *t*-Test**

	Unstandardized Coefficient	T	Significance
Constant	–.240	–2.939	0.019
T	0.044	33.999	0.000
lnKL	0.269	2.603	0.031
lnEL	0.407	4.296	0.003

That is, $c = 0.044$. $\alpha = 0.269$, $\beta = 1-\alpha-\gamma = 0.324$, $\gamma = 0.407$ $\alpha = 0.269$, which indicates that Y will increase 0.269% when K increases 1%; $\beta = 0.324$ indicates that Y will increase 0.324% when L increases 1%; $\gamma = 0.407$ indicates that Y will increase 0.407% when E increases 1%; $c = 0.044$ indicates a technological progress growth rate of 4.4%. According to (5.17), (5.3), and (5.5):

$$Y = 0,786628e^{0.044t} K^{0.269} L^{0.324} E^{0.407} \tag{5.18}$$

and

$$0.407 \frac{\dot{\chi}(t)}{\chi(t)} = 0.269 \times \frac{\dot{y}_k(t)}{y_k(t)} + 0.324 \times \frac{\dot{y}_l(t)}{y_l(t)} - 0.044 \tag{5.19}$$

Based on growth rate of technological progress of 0.044, we calculate economic growth as $(e^{0.044} - 1) \times 100\% = 4.4982\%$. We can now calculate the energy-saving effect of technological progress:

$$0.407 \frac{\dot{\chi}(t)}{\chi(t)} = 0.269 \times 0.044982 + 0.324 \times 0.044982 - 0.044 \tag{5.20}$$

The $(0.269 \times 0.44982 + 0.324 \times 0.044982)$ expression indicates that technological progress will promote economic and increase energy intensity. The subtraction of 0.044 at the end of the equation indicates that technological progress will decrease energy intensity. According to (5.20), we have

$$\frac{\dot{\chi}(t)}{\chi(t)}(T) = -4.26\%$$

where $\dot{\chi}(t)/\chi(t)(T)$ indicates the energy-saving effect of technological progress. In summary, technological progress will decreases the energy intensity of Chinese industry an average of 4.26% per year.

5.2.4 Conclusion

Capital, labor, and energy are treated as the elements of economic growth, and Cobb–Douglas production function results show that technological progress will decrease energy intensity. Empirical investigation of Chinese industry confirms that fact. ABWO was used to eliminate the interference of the 1997 Asian financial crisis to make the technological progress calculated by the Cobb–Douglas method more accurate.

5.3 Energy-Saving Effect of Industrial Restructuring

China is a developing country pursuing industrialization and modernization. Second industries account for a large proportion of the national economy. Since the mid-1990s, the share generated by primary industries has declined steadily. The proportion of secondary industries declined initially and then increased; tertiary industries followed the same pattern as shown in Figure 5.2. Secondary industries consume more energy, thus exerting negative effects on energy efficiency. We will

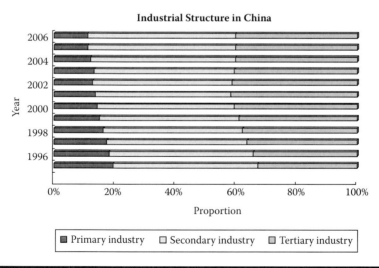

Figure 5.2 Industrial structure in China. (From *China Statistical Yearbook*, 2008. With permission.)

study the impact of China's industrial structure on energy consumption and use the grey linear programming model to analyze energy savings resulting from industrial restructuring.

5.3.1 Decomposition of Energy Intensity

Energy intensity indicates energy consumption per GDP:

$$e = E/Y$$

where E refers to energy consumption, and Y denotes the GDP. Energy consumption and GDP are decomposed according to the three categories of industries as follows:

$$E = \sum_{i=1}^{3} E_i = E_1 + E_2 + E_3 \tag{5.21}$$

$$Y = \sum_{i=1}^{3} Y_i = Y_1 + Y_2 + Y_3 \tag{5.22}$$

Energy intensity can be decomposed as:

$$e = \frac{E}{Y} = \frac{\sum_{i=1}^{3} E_i}{\sum_{i=1}^{3} Y_i} = \frac{\sum_{i=1}^{3} e_i Y_i}{\sum_{i=1}^{3} Y_i} = \sum_{i=1}^{3} e_i y_i \tag{5.23}$$

where e_i is the energy intensity of each type of industry, and y_i is the proportion of each type. Assume that e^n $(n = 1, 2,\ldots,N)$ indicates the energy intensity of period n and e^0 indicates the energy intensity of the base period; then:

$$e^n = \sum_{i=1}^{3} e_i^n y_i^n, \; e^0 = \sum_{i=1}^{3} e_i^0 y_i^0$$

and e^n can be decomposed:

$$e^n = \sum_{i=1}^{3} e_i^n y_i^n = \sum_{i=1}^{3} e_i^0 y_i^0 + \sum_{i=1}^{3} e_i^0 \left(y_i^n - y_i^0 \right) + \sum_{i=1}^{3} y_i^0 \left(e_i^n - e_i^0 \right) \tag{5.24}$$

The change of energy intensity can be decomposed as:

$$\Delta e = e^n - e^0 = \sum_{i=1}^{3} e_i^0 \left(y_i^n - y_i^0 \right) + \sum_{i=1}^{3} y_i^0 \left(e_i^n - e_i^0 \right) = \Delta e_s + \Delta e_e \quad (5.25)$$

where:

$\Delta e_s = \sum_{i=1}^{3} e_i^0 \left(y_i^n - y_i^0 \right)$ = change of energy intensity caused by industrial structure

$\Delta e_s = \sum_{i=1}^{3} y_i^0 \left(e_i^n - e_i^0 \right)$ = change of energy intensity caused by energy efficiency

5.3.2 Changes of Energy Intensity Caused by Industrial Structure

Based on value added and energy consumption statistics for all three types of industries, as shown in Table 5.8, the industrial structure changed from 2000 to 2007.

Based on energy intensities of the industry types as shown in Table 5.9 and using Equation (5.25), the change of energy intensity can be calculated as:

$$\Delta e_s = \sum_{i=1}^{3} e_i^0 \left(y_i^n - y_i^0 \right) = 0.404 \times 0.113 - 0.151 + 2.142 \times 0.486 - 0.459$$

$$+ 0.474 \times 0.401 - 0.39 = 0.0477$$

Table 5.8 Value Added and Energy Consumption of Chinese Industry in 2000 and 2007

	Value Added (10 Billion Yuan)[a]			Energy Consumption (Million tce)		
Year	Primary Industry	Secondary Industry	Tertiary Industry	Primary Industry	Secondary Industry	Tertiary Industry
2000	149.45	455.56	387.14	60.45	975.86	183.34
2007	197.36	971.93	732.61	82.45	1941.89	363.60

[a] Value added of each industry in 2007 was converted from the GDP index and calculated at 2000 constant prices.

Table 5.9 Chinese Industrial Structure and Energy Intensity

	Industrial Structure (%)			Energy Intensity (tce/10,000 Yuan)		
Year	Primary Industry	Secondary Industry	Tertiary Industry	Primary Industry	Secondary Industry	Tertiary Industry
2000	15.1	45.9	39.0	0.404	2.142	0.474
2007	11.3	48.6	40.1	0.418	1.998	0.496

Source: China Statistical Yearbook, 2008.

$$\Delta e_s = \sum_{i=1}^{3} y_i^0 \left(e_i^n - e_i^0 \right) = 0.113 \times \left(0.418 - 0.404 \right) + 0.486 \times \left(1.998 - 2.142 \right)$$

$$+ 0.401 \times \left(0.496305 - 0.473575 \right) = -0.0594$$

The improvement of energy efficiency decreased energy intensity 0.0594 tce per 10,000 yuan, but the industrial structure increased energy intensity 0.0477 tce per 10,000 yuan. The change of industrial structure increased the energy intensity between 2000 and 2007, indicating that industrial restructuring can help save energy.

5.3.3 Grey Linear Programming Model for Analyzing Industrial Restructuring Impact on Energy Saving

In this section, a grey linear programming model is created to study how to decrease energy intensity by industrial restructuring by analyzing fixed assets, labor, water resource, and other characteristics of economic growth.

The development of tertiary industry is closely related to that of the secondary industry in China, as shown in Table 5.10. Therefore, when restructuring the industry structure, we must consider this relationship. Secondary industries provide the necessary foundation for developing tertiary industries and tertiary industries provide service and security for developing secondary industries. To further analyze the relationship of China's secondary and tertiary industries, we used a linear regression model in which the value added of secondary industry is the independent variable and value added of tertiary industry is the dependent variable. The linear regression equation is:

$$Y_3 = \alpha + \beta Y_2 \tag{5.26}$$

where:
Y_3 = value added of tertiary industry
Y_2 = value added of secondary industry
Using SPSS for Equation (5.26):

$$Y_3 = -1596.847 + 0.858 Y_2 \tag{5.27}$$

The results of coefficients and related tests are shown in Tables 5.11 and 5.12. The adjusted R^2 of 0.989 and significant F change of 0.000 show that the regression model is effective. The t-test shows that Y_2 has a significant effect on Y_3. Regression equation results prove that the development of tertiary industry is closely related to development of secondary industry and must be a factor in industrial restructuring.

Table 5.10   Value Added of Secondary Industry and Tertiary Industry (10 Billion Yuan)

Year	Value Added of Secondary Industry	Value Added of Tertiary Industry
1991	91.02	73.37
1992	116.99	93.57
1993	164.54	119.16
1994	224.45	161.80
1995	286.80	199.79
1996	338.35	233.26
1997	375.43	269.88
1998	390.04	305.81
1999	410.33	338.73
2000	455.56	387.14
2001	495.12	443.62
2002	538.97	498.99
2003	624.36	560.05
2004	739.04	645.61
2005	873.65	734.33
2006	1031.62	847.21
2007	1213.81	1000.54

Source: China Statistical Yearbook, 2008.

Table 5.11   *R*-Test

			Change Statistics		
R	R^2	Adjusted R^2	R^2 Change	F Change	Significant F Change
0.995	0.990	0.989	0.990	1468.789	0.000

Table 5.12   Regression Coefficient and *t*-Test

Model		Unstandardized Coefficient		T	Significance
		B	Standard Error		
1	Constant	−1596.847	1301.119	−1.227	0.239
	Y2	0.858	0.022	38.325	0.000

By studying the real and simulated values of the regression model, as shown in Table 5.13, we find:

$$0.9Y_3 < -1596.847 + 0.858Y_2 < 1.1Y_3 \qquad (5.28)$$

Two constraints can be generated from equation (5.28):

$$-1596.847 + 0.858Y_2 < 1.1Y_3 \qquad (5.29)$$

$$0.9Y_3 < -1596.847 + 0.858Y_2 \qquad (5.30)$$

A linear programming model can be set up as follows:

$$\min \quad e_1Y_1 + e_2Y_2 + e_3Y_3$$

$$\begin{cases} a_{11}Y_1 + a_{12}Y_2 + a_{13}Y_3 \le b_1 \\ a_{21}Y_1 + a_{22}Y_2 + a_{23}Y_3 \le b_2 \\ a_{31}Y_1 + a_{32}Y_2 + a_{33}Y_3 \le b_3 \\ Y_1 + Y_2 + Y_3 \ge Y_A \\ -1596.847 + 0.858Y_2 < 1.1Y_3 \\ 0.9Y_3 < -1596.847 + 0.858Y_2 \\ Y_1 > 0 \\ Y_2 > 0 \\ Y_3 > 0 \end{cases} \qquad (5.31)$$

where:
e_1, e_2, e_3 = energy intensity of each industry
Y_1, Y_2, Y_3 = value added of each industry

Table 5.13 Relative Error of Regression Model

Year	Simulated Value $\hat{Y}_3$	Real Value Y_3	Residual $\hat{Y}_3 - Y_3$	Relative Error $(\hat{Y}_3 - Y_3/Y_3)$
1991	62.13	73.37	−11.24	−15%
1992	84.41	93.57	−9.16	−10%
1993	125.21	119.16	6.05	5%
1994	176.61	161.80	14.82	9%
1995	230.10	199.79	30.32	15%
1996	274.34	233.26	41.07	18%
1997	306.15	269.88	36.27	13%
1998	318.69	305.81	12.88	4%
1999	336.10	338.73	−2.63	−1%
2000	374.90	387.14	−12.24	−3%
2001	408.85	443.62	−34.77	−8%
2002	446.47	498.99	−52.52	−11%
2003	519.74	560.05	−40.31	−7%
2004	618.13	645.61	−27.48	−4%
2005	733.62	734.33	−0.71	0%
2006	869.16	847.21	21.95	3%
2007	1025.48	1000.54	24.95	2%

a_{11}, a_{12}, a_{13} = fixed assets per value added of each industry, which indicates necessary fixed assets of unit value added

a_{21}, a_{22}, a_{23} = labor per value added of each industry, which indicates necessary labor of unit value added

a_{31}, a_{32}, a_{33} = water resource per value added of each industry, which indicates water consumption of unit value added

b_1 = fixed assets

b_2 = labor

b_3 = water resource

YA = objective of regional economic growth

5.3.4 Industrial Restructuring

The programming model established in Section 5.3.3 was adopted to study China's industrial restructuring goals in 2010 and 2015. However, a variety of consumption coefficients (a_{ij}, $i, j = 1, 2, 3$), total resources, and the goals of economic development should be predicted before calculation of programming goals. This section adopts the GM (1,1) grey model to forecast these variables. Equation (5.31) is essentially a grey linear programming model; consumption coefficient ,total resources, and other parameters are grey.

The GDP at constant price for 2001 through 2007 (Table 5.14) is calculated from GDP indices based on an assumed GDP of 100 in 1978). Table 5.15 shows a GDP breakdown by industry for 2007. Table 5.16 shows GDP for 2001 through 2007 based on a constant price in 2007. The equation is:

$$Y_t = \frac{Y_{2007}}{I_{2007}} \times I_t \qquad (5.32)$$

Table 5.14   GDP, 2001–2007

Year	GDP	Primary Industry	Secondary Industry	Tertiary Industry
2001	823.0	284.8	1173.1	1054.2
2002	897.8	293.0	1288.4	1164.2
2003	987.8	300.3	1451.7	1274.9
2004	1087.4	319.3	1613.0	1403.1
2005	1200.8	336.0	1801.6	1550.4
2006	1340.7	352.8	2035.2	1738.1
2007	1500.7	365.8	2307.7	1956.3

Source: China Statistical Yearbook, 2008.

Note: Assume 1978 GDP is 100.

Table 5.15   GDP in 2007 (10 Billion Yuan)

GDP	Primary Industry	Secondary Industry	Tertiary Industry
2495.30	280.95	1213.818	1000.54

Source: China Statistical Yearbook, 2008.

Table 5.16 GDP (Constant Price in 2007; 10 Billion Yuan)

Year	GDP	Primary Industry	Secondary Industry	Tertiary Industry
2001	1368.49	218.71	617.03	539.14
2002	1492.77	225.05	677.68	595.44
2003	1642.43	230.67	763.56	652.02
2004	1808.07	245.21	848.41	717.59
2005	1996.70	258.03	947.63	792.94
2006	2229.26	270.93	1070.50	888.94
2007	2495.30	280.95	1213.81	1000.54

Table 5.17 Industry Employment

Year	Primary Industry	Secondary Industry	Tertiary Industry	Employment (Million Persons)
2001	365.13	162.84	202.28	730.25
2002	368.70	157.80	210.90	737.40
2003	365.46	160.77	218.09	744.32
2004	352.69	169.20	230.11	752.00
2005	339.70	180.84	237.71	758.25
2006	325.61	192.25	246.14	764.00
2007	314.44	206.29	249.17	769.90

Source: China Statistical Yearbook, 2008.

where:

Y_t = value added or GDP in year t

Y_{2007} = value added or GDP in 2007

I_{2007} = the index of value added or index of GDP in 2007

I_t = the index of value added or index of year t

The GDP calculated at constant price in 2007 is shown in Table 5.16. Total employment by industry is shown in Table 5.17; water consumption by industry is shown in Table 5.18.

Although capital stock is an important macroeconomic variables required to study economic growth and total factor productivity (TFP), the present economic accounting system currently does not contain capital stock data. The stock of fixed

Table 5.18 Industry Water Use (100 Million Cubic Meters), 2001–2007

Year	Total Water Consumption	Primary Industry	Secondary Industry	Tertiary Industry
2001	5567.4	3825.7	1141.8	599.9
2002	5497.3	3736.2	1142.4	618.7
2003	5240.9	3432.8	1177.2	630.9
2004	5465.8	3585.7	1228.9	651.2
2005	5540.3	3580.0	1285.2	675.1
2006	5702.0	3664.4	1343.8	693.8
2007	5712.9	3599.5	1403.0	710.4

Source: China Statistical Yearbook, 2008.

capital is estimated based on depreciation data. The advantage of this approach is that only one variable must be estimated. The rate of depreciation is:

$$FA_{t-1} = DFA_t / \sigma \tag{5.33}$$

where:
FA_{t-1} = stock of fixed capital during year $t-1$
DFA_t = depreciation of fixed assets in year t
σ = rate of depreciation of fixed assets

The rate of depreciation was set at 5% in a study of Wang Xiaolu and Fan Gang (2000). Stock of fixed capital was estimated according to the rate of depreciation calculated by Equation (5.33). Depreciation of fixed assets by region was published in the *China Statistical Yearbook*, as shown in Table 5.19. The results of capital stock estimation are shown in Table 5.20.

To calculate the capital consumption coefficient of each type of industry, the stock of fixed assets had to be decomposed by industry. The growth of industrial value added relates to fixed assets closely. The growth rate of fixed assets is replaced by value added for estimating the stock of fixed assets. The distribution proportion of fixed assets is as follows:

$$y_t^j = \frac{y_{t-1}^j \left(1 + g_t^j\right)}{\sum\limits_{j=1}^{3} y_{t-1}^j \left(1 + g_t^j\right)} \quad (j = 1, 2, 3) \tag{5.34}$$

Table 5.19 Depreciation of Fixed Assets of Provinces (100 Million Yuan)

Region	2004	2005	2006	2007
Beijing	573.08	1095.51	1251.09	1375.46
Tianjin	440.20	515.77	595.09	754.32
Hebei	1012.98	1254.97	1491.98	1817.05
Shanxi	373.38	634.89	698.24	816.82
Inner Mongolia	274.71	544.87	701.28	969.23
Liaoning	1088.27	1288.09	1490.67	1688.23
Jilin	419.98	617.51	766.42	907.38
Heilongjiang	708.91	733.35	809.93	960.90
Shanghai	941.10	1501.79	1730.51	1951.84
Jiangsu	1907.20	2960.06	3232.53	3582.72
Zhejiang	1152.81	1905.37	2191.19	2593.49
Anhui	555.46	738.25	811.63	997.51
Fujian	745.23	919.33	1002.02	1097.12
Jiangxi	569.71	488.88	541.90	618.80
Shandong	2701.85	2940.02	3271.02	3771.38
Henan	984.51	1220.15	1339.45	1579.84
Hubei	890.28	1098.99	1172.59	1386.03
Hunan	717.99	847.17	990.76	1188.08
Guangdong	2073.56	3585.19	4146.68	4558.87
Guangxi	339.38	496.15	522.95	643.96
Hainan	108.37	142.46	162.54	235.81
Chongqing	279.55	376.54	436.48	432.76
Sichuan	920.20	1246.35	1469.67	1788.94
Guizhou	172.53	290.33	346.98	380.85
Yunnan	385.56	512.76	577.08	645.28
Xizang	36.31	57.56	63.28	74.32

(Continued)

Table 5.19 (Continued)   Depreciation of Fixed Assets of Provinces (100 Million Yuan)

Region	2004	2005	2006	2007
Shanxi	438.28	610.93	762.31	912.20
Gansu	274.10	300.29	385.60	467.18
Qinghai	68.39	102.17	121.36	149.43
Ningxia	79.82	122.71	133.65	178.36
Xinjiang	317.77	373.58	424.96	494.69

Source: China Statistical Yearbook, 2008.

Table 5.20   Estimated Capital Stock

Year	Depreciation of Fixed Assets (10 Billion Yuan)	Estimated Capital Stock (10 Billion Yuan)	Price Index of Fixed Asset Investments	Estimated Capital Stock (10 Billion Yuan)*
2003	–	4310.29	186.4	4874.52
2004	215.51	5904.40	196.8	6324.43
2005	295.22	6728.39	199.9	7095.25
2006	336.42	7803.77	202.9	8107.61
2007	390.19	–	210.8	–

*Constant price in 2007.

where:

y_{t-1}^{j} = proportion of capital stock of industry j in year $t-1$

y_{t}^{j} = proportion of capital stock of industry j in year t

g_{t}^{j} = growth rate of industry j calculated via index of GDP (Table 5.21)

Distribution of fixed assets by base year is needed for calculating the distribution of capital stock. Using Equation (5.34), Xiang Xu estimated the capital stock of the three types of industries in the provinces of China. Due to data limitations, the estimated results only extended 2002 and capital stock was calculated based on constant prices from 1978. By analyzing those results, we found that the proportions of stock of fixed assets of each industry type in 2002 were 0.0426, 0.4544, and 0.5030, respectively. Distributions for subsequent years based on Equation (5.34) are shown in Table 5.22. Estimates of fixed assets by industry are shown in Table 5.23.

Table 5.21   GDP Index (100 in 2002)

Year	Primary Industry	Secondary Industry	Tertiary Industry
2003	102.5	112.7	109.5
2004	106.3	111.1	110.1
2005	105.2	111.7	110.5
2006	105.0	113.0	112.1
2007	103.7	113.4	112.6

Table 5.22   Distribution Proportions of Fixed Assets

Year	Primary Industry	Secondary Industry	Tertiary Industry
2003	0.0395	0.4629	0.4976
2004	0.0380	0.4657	0.4963
2005	0.0360	0.4693	0.4947
2006	0.0337	0.4724	0.4939

Table 5.23   Fixed Assets in Industry (10 Billion Yuan)

Year	Primary Industry	Secondary Industry	Tertiary Industry
2003	192.30	2256.24	2425.97
2004	240.20	2945.58	3138.64
2005	255.73	3329.69	3509.82
2006	273.30	3829.64	4004.67

The coefficient of resource consumption in each industry can be calculated based on consumption; the labor input coefficient is shown in Table 5.24, the water consumption coefficient is shown in Table 5.25, and the fixed assets consumption coefficients are shown in Table 5.26.

We predicted the totals of all resource and consumption coefficients according to the grey model GM (1,1). The total employment forecasting time-responsive equation is:

$$\hat{x}^{(1)}(k+1) = 8524101.830458e^{0.008622k} + 8451076.830458$$

Table 5.24   Labor Input Coefficients (10,000 Persons per 100 Million Yuan)

Year	Primary Industry	Secondary Industry	Tertiary Industry
2001	1.6695	0.2639	0.3752
2002	1.6383	0.2329	0.3542
2003	1.5843	0.2106	0.3345
2004	1.4383	0.1994	0.3207
2005	1.3165	0.1908	0.2998
2006	1.2018	0.1796	0.2769
2007	1.1192	0.1699	0.2490

Table 5.25   Water Consumption Coefficients (100 Million Cubic Meters per 100 Million Yuan)

Year	Primary Industry	Secondary Industry	Tertiary Industry
2001	0.1749	0.0185	0.0111
2002	0.1660	0.0169	0.0104
2003	0.1488	0.0154	0.0097
2004	0.1462	0.0145	0.0091
2005	0.1387	0.0136	0.0085
2006	0.1353	0.0126	0.0078
2007	0.1281	0.0116	0.0071

Table 5.26   Consumption of Fixed Capital Coefficients

Year	Primary Industry	Secondary Industry	Tertiary Industry
2003	0.8337	2.9549	3.7207
2004	0.9796	3.4719	4.3738
2005	0.9910	3.5136	4.4263
2006	1.0087	3.5774	4.5050

The simulated values of total employment and the relative error are shown in Table 5.27.

The average relative error was calculated as0.07835% and is very accurate. The total employment estimates for 2010 and 2015 are shown in Table 5.28.

The total water supplied forecasting time-responsive equation is

$$\hat{x}^{(1)}(k+1) = 402121.415608e^{0.013206k} + 396554.015608$$

The simulated values of water supplied and relative error are shown in Table 5.29.

Table 5.27 Total Employment Forecast Errors

Simulated Value	Absolute Error	Relative Error (%)
738.14	74.15	0.10
744.53	21.34	0.03
750.98	−101.93	−0.14
757.48	−76.62	−0.10
764.04	4.32	0.01
770.66	75.94	0.10

Table 5.28 Total Employment Estimate (Million Persons)

Year	2010	2015
Estimate	790.85	825.69

Table 5.29 Water Supplied Forecast Error

Simulated Value	Absolute Error	Relative Error (%)
5345.59	−151.71	−2.76
5416.66	175.76	3.35
5488.66	22.86	.418
5561.63	21.33	.385
5635.56	−66.44	−1.17
5710.48	−2.42	−.0424

The average relative error is 1.354% and is very accurate. Total water supplied estimates for 2010 and 2015 are shown in Table 5.30.

The stock of fixed assets forecasting time-responsive equation is

$$\hat{x}^{(1)}(k+1) = 4733379.698236e^{0.124789k} - 4245928.196438$$

The simulated values of the stock of fixed assets and the relative error are shown in Table 5.31.

The calculated average relative error of 0.462% is very accurate. The fixed asset estimates for 2010 and 2015 are shown in Table 5.32.

The water consumption coefficient forecasting time-responsive equation for each type of industry is:

Primary industry: $\hat{x}^{(1)}(k+1) = -3.464431e^{-0.047747k} + 3.639356$
Secondary industry: $\hat{x}^{(1)}(k+1) = -0.23721e^{-0.073296k} + 0.255715$
Tertiary industry: $\hat{x}^{(1)}(k+1) = -0.147397e^{-0.073471k} + 0.158524$

The simulated values and the relative error are shown in Table 5.33.

The calculated average relative errors of 1.5844, 0.558, and 0.864% are very accurate. Water consumption coefficient estimates for all three types of industries for 2010 and 2015 are shown in Table 5.34.

**Table 5.30   Total Water Supplied
Estimate (Billion Cubic Meters)**

Year	2010	2015
Estimate	594.13	634.68

Table 5.31   Stock of Fixed Assets Forecast Error

Simulated Value	Absolute Error	Relative Error (%)
629111.97	−3330.66	−0.527
712727.03	3202.28	0.451
807455.35	−3305.96	−0.408

**Table 5.32   Fixed Assets Estimate
(Hundred Billion Yuan)**

Year	2010	2015
Estimate	1174.10	2191.19

Table 5.33 Water Consumption Coefficient Forecast Error

Primary Industry			Secondary Industry			Tertiary Industry		
Simulated Value	Absolute Error	Relative Error (%)	Simulated Value	Absolute Error	Relative Error (%)	Simulated Value	Absolute Error	Relative Error (%)
0.16	0.00	−2.70	0.02	0.00	−0.55	0.01	0.00	0.48
0.15	0.01	3.48	0.02	0.00	1.06	0.01	0.00	0.27
0.15	0.00	0.40	0.01	0.00	−0.04	0.01	0.00	−0.67
0.14	0.00	0.89	0.01	0.00	−0.79	0.01	0.00	−1.62
0.13	0.00	−1.33	0.01	0.00	−0.38	0.01	0.00	−0.29
0.13	0.00	−0.70	0.01	0.00	0.53	0.01	0.00	1.85

Table 5.34   Water Consumption Coefficient Estimate (100 Million Cubic Meters per 100 Million Yuan)

Year	Primary Industry	Secondary Industry	Tertiary Industry
2010	11.02	0.93	0.58
2015	8.68	0.65	0.40

The labor consumption coefficient forecasting time-responsive equation for each type of industry is as follows:

Primary industry: $\hat{x}^{(1)}(k+1) = -21.86834e^{-0.079501k} + 23.537849$
Secondary industry: $\hat{x}^{(1)}(k+1) = -3.864271e^{-0.060902k} + 4.128179$
Tertiary industry: $\hat{x}^{(1)}(k+1) = -5.582465e^{-0.066415k} + 5.957655$

The simulated values of labor consumption coefficient and relative error are shown in Table 5.35.

The calculated average relative errors of 1.168, 1.157, and 1.609% are very accurate. Labor consumption coefficient estimates for 2010 and 2015 for all types of industries are shown in Table 5.36.

The fixed assets consumption coefficient forecasting time-responsive equation is:

Primary industry: $\hat{x}^{(1)}(k+1) = 66.13792e^{0.014688k} - 65.304254$
Secondary industry: $\hat{x}^{(1)}(k+1) = 229.460519e^{0.015002k} - 226.505613$
Tertiary industry: $\hat{x}^{(1)}(k+1) = 293.0315e^{0.014801k} - 289.310828$

The simulated values of fixed assets consumption coefficient and relative error by industry type are shown in Table 5.37.

The average relative errors calculated as 0.134, 0.134, and 0.127% are very accurate. The fixed assets consumption coefficient estimates by industry for 2010 and 2015 are shown in Table 5.38.

Assuming an 8% economic growth rate, the GDP of China should have reached 314.34 (100 billion yuan) in 2010 and should reach at least 461.86 (100 billion yuan) in 2015. The programming models for 2010 (5.35) and 2015 (5.36) can be obtained according to Equation (5.31).

Table 5.35 Labor Consumption Coefficient Forecast Error

Primary Industry			Secondary Industry			Tertiary Industry		
Simulated Value	Absolute Error	Relative Error (%)	Simulated Value	Absolute Error	Relative Error (%)	Simulated Value	Absolute Error	Relative Error (%)
1.67	0.03	2.01	0.23	0.00	−1.95	0.36	0.00	1.28
1.54	−0.04	−2.57	0.21	0.00	2.03	0.34	0.00	0.35
1.43	−0.01	−0.89	0.20	0.00	1.36	0.31	−0.01	−2.05
1.32	0.00	0.01	0.19	0.00	−0.33	0.29	−0.01	−1.96
1.22	0.01	1.18	0.18	0.00	−0.35	0.28	0.00	−0.67
1.12	0.00	0.35	0.17	0.00	−0.92	0.26	0.01	3.34

Table 5.36 Industry Labor Consumption Coefficient Estimate (10,000 Persons per 100 Million Yuan)

Year	Primary Industry	Secondary Industry	Tertiary Industry
2010	0.885	0.140	0.211
2015	0.595	0.103	0.151

$$Min\ 0.4178Y^1_{2010} + 1.9980Y^2_{2010} + 0.4963Y^3_{2010}$$

$$
\begin{cases}
1.0532Y^1_{2010} + 3.7884Y^2_{2010} + 4.7051Y^3_{2010} \leq 1174 \\[4pt]
0.8847Y^1_{2010} + 0.1402Y^2_{2010} + 0.2108Y^3_{2010} \leq 790 \\[4pt]
0.1102Y^1_{2010} + 0.0093Y^2_{2010} + 0.0058Y^3_{2010} \leq 594 \\[4pt]
\quad Y^1_{2010} + Y^2_{2010} + Y^3_{2010} \geq 3143 \\[4pt]
\quad -15.96 + 0.858Y^2_{2010} < 1.1Y^3_{2010} \\[4pt]
\quad 0.9Y^3_{2010} < -15.96 + 0.858Y^2_{2010} \\[4pt]
\qquad Y^1_{2010} > 0 \\[4pt]
\qquad Y^2_{2010} > 0 \\[4pt]
\qquad Y^3_{2010} > 0
\end{cases}
\tag{5.35}
$$

and

$$Min\ 0.4178Y^1_{2015} + 1.9980Y^2_{2015} + 0.4963Y^3_{2015}$$

$$
\begin{cases}
1.1334Y^1_{2015} + 4.0295Y^2_{2015} + 5.0665Y^3_{2015} \leq 2191 \\[4pt]
0.5945Y^1_{2015} + 0.1034Y^2_{2015} + 0.1513Y^3_{2015} \leq 825 \\[4pt]
0.0868Y^1_{2015} + 0.0065Y^2_{2015} + 0.0040Y^3_{2015} \leq 634 \\[4pt]
\quad Y^1_{2015} + Y^2_{2015} + Y^3_{2015} \geq 4618 \\[4pt]
\quad -15.96 + 0.858Y^2_{2015} < 1.1Y^3_{2015} \\[4pt]
\quad 0.9Y^3_{2015} < -15.96 + 0.858Y^2_{2015} \\[4pt]
\qquad Y^1_{2015} > 0 \\[4pt]
\qquad Y^2_{2015} > 0 \\[4pt]
\qquad Y^3_{2015} > 0
\end{cases}
\tag{5.36}
$$

Table 5.37 Fixed Assets Consumption Coefficient Forecast Error

Primary Industry			Secondary Industry			Tertiary Industry		
Simulated Value	Absolute Error	Relative Error (%)	Simulated Value	Absolute Error	Relative Error (%)	Simulated Value	Absolute Error	Relative Error (%)
0.9790	−0.0010	−0.1049	3.4680	−0.0036	−0.1049	4.3690	−0.0044	−0.0994
0.9930	0.0020	0.1992	3.5210	0.0070	0.1985	4.4350	0.0083	0.1881
1.0080	−0.0010	−0.0992	3.5740	−0.0035	−0.0989	4.5010	−0.0042	−0.0939

Table 5.38   Fixed Assets Consumption Coefficient Estimate

Year	Primary Industry	Secondary Industry	Tertiary Industry
2010	1.05316	3.788398	4.705126
2015	1.133413	4.029594	5.066544

Table 5.39   Value Added, Proportion, and Energy-Saving Effects

		2010	2015
Value added (10 billion yuan)	Primary industry	340.83	479.42
	Secondary industry	1582.61	2128.13
	Tertiary industry	1219.92	2011.08
Industry structure (%)	Primary industry	10.8	10.4
	Secondary industry	50.3	46.1
	Tertiary industry	38.8	43.5
Energy-saving effect (tce/10,000 yuan)		0.02661	–0.03716

Note: Value added is calculated at constant price in 2007; energy-saving effect calculated on base year 2007.

where Y_t^j ($j = 1, 2, 3; t = 2010, 2015$) is the value added of industry j in year t.

The value added in 2010 and 2015 can be calculated by solving the programming model (5.35) and (5.36). We use the industry structure based on value added; the energy-saving effects of industry restructuring can be calculated according to Equation (5.25). The results appear in Table 5.39.

The result of industrial restructuring for 2010 shows that the proportion of secondary industries increases and tertiary industries are declining because China is focusing on industrialization. The increased proportion of secondary industries is important; greater levels of industrialization will allow tertiary industries to develop faster. The proportion of secondary industries will decrease and tertiary industries will increase more in 2015 than in 2010; we assumed that the proportion of tertiary industries in 2015 would increase over the level for 2010- by nearly 5 percentage points. This change of industrial structure will increase the energy intensity in the short term, and then reduce the intensity.

5.3.5 Conclusion

In this section, we considered capital, water, labor, and other constraints to establish a grey linear programming model to study the impact of industrial structure adjustment on energy consumption and determine how to ensure fast and steady development of the economy. We found that a change of China's industrial structure will increase the energy consumption per unit GDP to some extent for a few years and then decrease the consumption. The industrial structure adjustment is very important for China's energy consumption in the future but it will also exert greater pressure on energy consumption.

5.4 Energy-Saving Effect of Development and Use of Nonfossil Energy

China's energy consumption is dominated by coal—about 70% of total energy consumption for a long time, as shown in Table 5.40. Research indicates that coal combustion in China releases 85% of the sulfur dioxide, 70% of the dust, 85% of the carbon dioxide, and 60% of the nitrogen oxides into the atmosphere. Unlike oil, gas, and other relatively clean energy sources, coal is *dirty* energy. China's low efficiency coal consumption is from the mainstream trend of world energy use.

Chinese energy consumption is influenced by exploitation of its reserves. Coal is much easier to obtain than oil and natural gas and coal production supported the rapid growth of Chinese energy consumption. Considering the keen international competition for oil and gas resources, it will be difficult for China to adjust its energy structure to compete with developed countries that already entered the oil age. Nuclear, hydro, wind, and other nonfossil energies will become important in China. Most of these nonfossil energy sources also save energy. This section

Table 5.40 Energy Consumption Structure (%)

Year	Coal	Oil	Gas	Hydro, Nuclear, and Wind Power
2000	67.8	23.2	2.4	6.7
2001	66.7	22.9	2.6	7.9
2002	66.3	23.4	2.6	7.7
2003	68.4	22.2	2.6	6.8
2004	68.0	22.3	2.6	7.1
2005	69.1	21.0	2.8	7.1
2006	69.4	20.4	3.0	7.2
2007	69.5	19.7	3.5	7.3

Source: China Statistical Yearbook, 2008.

examines structural changes in China's energy consumption trends and forecast changes in its energy consumption in the future.

5.4.1 Energy Consumption Structure

A quadratic programming model of the energy structure based on the homogeneous Markov chain utilized coal, oil, natural gas, and nonfossil (hydro, nuclear, and wind) energy consumption structure data for 2000 through 2007 to determine energy consumption trends. The future structure of energy consumption is related only to the current structure of energy consumption and has no bearing on the past so Markov chain model is appropriate for analysis. The model is:

$$W_j(t) = \sum_{i=1}^{4} W_i(t-1) P_{ij} + \varepsilon_j(t) \tag{5.37}$$

where:

$i, j = 1, 2, 3, 4;$ = coal, oil, natural gas, and nonfossil energy, respectively
$W_j(t)$ = proportion of j-th energy consumption at t time
$\varepsilon_j(t)$ = random interference
P_{ij} = transition probability (probability that i-th energy was consumed at $t-1$ time while j-th energy was consumed at time t)

To ensure that P_{ij} can correctly reflect the trend of Chinese energy consumption, the average transition probability P_{ij} must minimize $\Sigma_{t=2}^{n} \Sigma_{j=2}^{4} \varepsilon_j(t)$. The following quadratic programming model is applied to solve average transition probability P_{ij}.

$$\min \sum_{t} \sum_{j} \left(W_j(t) - \sum_{i=1}^{4} W_i(t-1) P_{ij} \right)^2$$

$$s.t \begin{cases} \sum_{i=1}^{4} P_{ij} = 1 \\ P_{ij} \geq 0 \end{cases} \tag{5.38}$$

The data for Chinese energy consumption from 2000 to 2007 are shown in Table 5.40. Nonfossil energy (nuclear, hydro, and wind) will not be replaced by the other energies, so $p_{44} = 1$. Natural gas resources are undergoing rapid development. Coal, oil, and nonfossil energy will not replace natural gas, so $p_{33} = 1$ and we add $p_{33} = 1$ and $p_{44} = 1$ to the bounds of (5.38). The quadratic programming model is solved by Excel. The results are shown in Table 5.41.

Table 5.41   Average Transition Probability Matrix for Energy

	Coal	*Oil*	*Natural Gas*	*Nonfossil Energy*
Coal	0.987	0	0	0
Oil	0	0.989	0	0
Natural gas	0	0.011	1	0
Nonfossil energy	0.013	0	0	1
Total	1	1	1	1

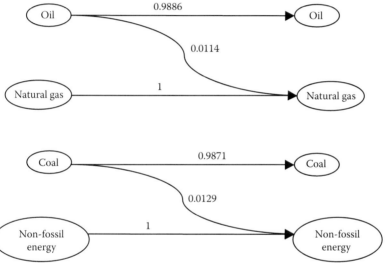

Figure 5.3   Chinese energy consumption trend.

The average transition probability *Pij* is depicted in Figure 5.3. Note that two major trends will change the energy consumption structure: (1) partial replacement of coal by nonfossil energy; and (2) partial replacement of oil by natural gas.

5.4.2 Energy Consumption Structure Forecasting

China's energy consumption structure forecast based on Equation (5.37) is shown in Table 5.42. One change of trend is the decline of coal use that will improve the unreasonable energy consumption structure to an extent. The proportion of oil will decline slightly—China will not enter the oil to a great degree. Natural gas consumption will grow faster so the proportion of natural gas will increase. The future

Table 5.42   Energy Consumption Structure Forecasting

Year	2009	2010	2015
Coal	67.7%	66.8%	62.6%
Oil	19.3%	19.0%	18.0%
Natural gas	3.9%	4.2%	5.2%
Non-fossil energy	9.1%	10.0%	14.2%

for nonfossil energy consumption is optimistic, and the proportion of nonfossil energy will increase greatly. China's energy consumption structure prediction is in line with the trend of its energy development.

5.4.3 Conclusion

A quadratic programming model was used to study the changing trend of China's energy consumption structure. The Markov was suitable for predicting energy consumption in 2010 and 2015. The results show that the proportion of coal will decline and the proportion of oil will decrease slightly. The proportion of natural gas and nonfossil energy sources will increase a little. Combined with the National Development and Reform Commission's energy-related planning, The forecast appears consistent with China's energy production and consumption trends. From a broader perspective, the development of nonfossil energy will aid energy conservation. The results in this section show that adjustments of energy consumption, particularly the development of nonfossil energy sources, will reduce China's overall consumption level.

5.5 Evaluation of Energy Policy

Since the 1980s, the Chinese government has implemented many energy policies; the milestones are shown in Table 5.43. Because of the overlaps of energy policies, it is difficult to evaluate the effects of individual policies. In this section, we classify the policies into three groups according to the relationships of principal and subordinate policies to better assess the effects of each group of energy policies.

In 1991, the State Planning Commission promulgated several suggestions to enhance energy saving:

1. The Economic and Trade Committee is responsible for supervision of national energy-saving programs and is responsible for training energy management staffs of key energy consumption units on energy saving.

Table 5.43   History of Energy-Saving Policies

Year	Policy
1986	Interim regulation of energy-saving management
1990	Energy-saving provision in eighth 5-year plan
1991	Regulation of energy saving for thermal power plants
1991	Proposals to enhance energy savings
1991	Regulation of grading and upgrading of energy-saving measures by commercial enterprise
1992	Suggestions for innovative housing materials and energy-saving architecture
1994	Suggestions to enhance saving and utilization of natural resources
1996	Regulation of technological innovation projects to save energy
1995	Energy-saving provision in ninth 5-year plan
1996	Regulation of supervision of energy saving by ministry responsible for coal industry
1997	Energy-saving design standards for civil construction
1998	Law requiring energy conservation
1999	Regulation of key energy-using units
2000	Regulation of energy-saving for civil construction
2001	Regulation for saving electricity
2004	Medium and long-term schema to save energy
2005	Implementation of design standards for energy saving in new civil construction
2005	Promotion of energy and land saving measures for civil and public construction
2005	Suggestions promoting innovative housing materials and energy-saving architecture
2005	Regulation of energy-saving in civil construction, abolishing 2000 regulation
2006	Implementation of energy-saving plan for 1,000 enterprises

(Continued)

Table 5.43 (Continued)   History of Energy-Saving Policies

Year	Policy
2006	Prohibition of blind reexpansion of high energy consuming industries
2006	Reinforcement of energy-saving issues
2006	Administrative reply to plan to reduce energy intensity indices of all regions during 11th 5-year plan
2006	Suggestion to implement top ten key energy-saving projects in 11th 5-year plan
2007	Suggestions for energy saving and emission reduction in coal industry
2007	Comprehensive scheme to save energy and reduce pollutant emissions
2007	Reinforcement of industrial structure adjustment to prevent blind reexpansion of high energy consuming industries
2007	Energy development plan for 11th 5-year plan
2007	Medium- and long-term development plan for renewable energy
2007	Comprehensive energy reduction program
2007	Energy-saving statistics, monitoring, assessment programs, and approaches to implement
2007	Amending energy-saving law

2. Key energy consumption units should implement the state's energy laws, regulations, guidelines, policies, and standards and accept supervision and inspection by the Economic and Trade Commission. Key energy consumption units should establish sound energy management systems using scientific management methods and advanced technical means to develop and implement conservation plans and measures for using energy reasonably and efficiently; allocate funds annually for energy research and development, transformation, promotion of energy-saving practices, and training; establish sound measurement, monitoring, and management systems and buy qualified energy measurement equipment; monitor energy consumption and energy use reporting systems and designate a person responsible for maintaining statistics. They should establish and improve the original records, establish energy consumption cost management systems and develop advanced and reasonable consumption limits; implement energy cost management; create energy

conservation responsibility systems; reduce consumption, improve economic efficiency; and create energy management positions.

3. Governments should recognize and reward key energy consumption units and individuals who make significant contributions to energy management and saving. Key energy units should allocate funds for rewarding groups and individuals who achieve energy savings and ensure that their energy policies comply with regulations.

The law covering energy conservation amended and adopted on October 28, 2007 was first promulgated in 1998. This shows that China started to regulate energy management before the new millennium.

In 2004, the National Development and Reform Commission compiled a "Medium and Long Term Specific Schema on Energy Saving," detailing development goals and priorities through 2020. The five sections covered (1) China's energy consumption situation; (2) conservation tasks; (3) guidance principles and objectives for energy conservation; (4) focus areas and key projects of energy conservation; and (5) environmental protection measures. The schema emphasized the adjustment of China's industrial structure to achieve energy saving, technological progress, and sustainable development. It cites the connection between energy saving and economic development as a critical factor. The 1991 suggestions to further reinforce energy saving, the 1998 law on energy conservation, and the 2004 medium and long term specific schema represent three different energy-saving phases promoted by the Chinese government.

5.5.1 Model

In this section, we apply two models to evaluate the energy-saving effects of energy policies: (1) with and without antitheses and (2) linear regression.

5.5.1.1 With and without Antitheses

This method is effective because all the factors except energy policies are stable. The first step is to collect the data on energy intensity including historical data before implementing the energy policies and real values resulting from implementing of the policies. The second step is to predict future energy intensity according to the historical data and calculated values for the energy intensity without the effects of the policies. The predicted values indicate the energy intensity under the hypothesis that the influences of all factors are stable. This study required a long time span and analysis of many influencing factors. Older data cannot reflect recent tendencies; only new data based on recent tendencies is appropriate to use for predicting energy intensity. The GM (1,1) is an important grey system method that is useful for small samples (Liu and Lin, 1998). Predicted values can be obtained by GM(1,1) with newer data and can reflect the trend of energy intensity more accurately. The third

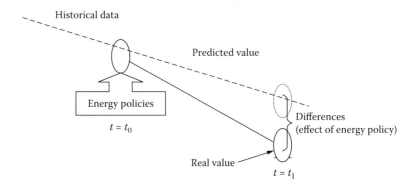

Figure 5.4   With and without antitheses.

step is to calculate the differences between real (after implementing energy policies) and predicted values of energy intensity. The differences represent the effects of energy policies. Figure 5.4 depicts the principle.

5.5.1.2 Linear Regression

Many researchers studied Chinese energy intensity. The main factors affecting energy intensity in China are economic growth, economic structure, energy prices, technological progress, and energy policies. GDP indicates the influence of energy consumption caused by economies of scale. Secondary and tertiary industry proportions are the factors indicating economic structure; and total factor productivity is used to evaluate technological progress. A linear regression model is created,

$$EI = \alpha + \beta_1 GDP + \beta_2 EP + \beta_3 TP + \beta_4 SP + \beta_5 TFP + \beta_6 P_1 + \beta_7 P_2 + \beta_8 P_3 + u_t \quad (5.39)$$

where:
EI = energy intensity
GDP = gross domestic production
EP = energy price
SP = secondary industry proportion
TP = tertiary industry proportion
TFP = total factor productivity
P_1, P_2, P_3 = groups of energy policies
ut = perturbation

5.5.2 Data

Energy consumption, energy price indices, GDP indices, secondary and tertiary industry proportions, and TFP are shown in Table 5.44. Because of the lack of

Table 5.44 Energy Consumption, Indices Energy Price and GDP, Proportions of Secondary and Tertiary Industry, and TFP

Year	Energy Consumption	Index of Energy Price	Index of GDP	Proportion of Secondary Industry	Proportion of Tertiary Industry	TFP
1982	62646	100.5	133.1	44.8	21.8	104.2
1983	66040	106.3	147.6	44.4	22.4	105.4
1984	70904	112.0	170.0	43.1	24.8	106.6
1985	76682	107.2	192.9	42.9	28.7	107.9
1986	80850	104.6	210.0	43.7	29.1	109.3
1987	86632	104.0	234.3	43.6	29.6	110.4
1988	92997	106.8	260.7	43.8	30.5	111.4
1989	96934	108.4	271.3	42.8	32.1	112.0
1990	98703	107.1	281.7	41.3	31.6	111.3
1991	103783	118.8	307.6	41.8	33.7	111.9
1992	109170	115.3	351.4	43.4	34.8	112.8
1993	115993	171.3	400.4	46.6	33.7	113.9
1994	122737	148.7	452.8	46.6	33.6	115.0
1995	131176	121.2	502.3	47.2	32.9	116.1
1996	138948	104.6	552.6	47.5	32.8	117.1
1997	137798	107.4	603.9	47.5	34.2	118.0
1998	132214	93.0	651.2	46.2	36.2	119.1
1999	133831	109.6	700.9	45.8	37.7	120.1
2000	138553	144.3	759.9	45.9	39.0	121.0
2001	143199	99.1	823.0	45.1	40.5	121.7
2002	151797	95.2	897.8	44.8	41.5	122.7
2003	174990	115.6	987.8	46.0	41.2	123.8
2004	203227	114.2	1087.4	46.2	40.4	124.8
2005	224682	122.4	1200.8	47.5	40.0	125.8
2006	246270	120.3	1334.0	48.9	39.4	126.8

Source: China Statistical Yearbook, 1996–2007.

energy price indices and the high market prices of oil in China, factory price indices of the petroleum industry were used instead of energy price indices. In 2006, the GDP of China was 211.81 (1,000) trillion yuan and the GDP index was 1334.0. The real GDP was calculated as 211808.0 divided by 1334. Energy intensity was calculated as energy consumption divided by real GDP. We assumed the energy price in 1981 was 1.0; energy price was defined as the product of energy price indices.

TFP data is from the research of Yang Yang (2008). The study used latent variables adopted to measure TFP. This approach characterized TFP as an independent variable to rule out the influences of other factors. TFP calculated via this approach is suitable for measuring technological progress.

Dummy variables P1, P2, and P3 indicate energy policies. P1 indicates energy policy group 1. P1 values were set at 0 before 1991, 1 from 1991 to 1997 when the policy was promulgated, and 0 after 1998. After the law on energy conservation, energy management in China changed and the influence of P1 became less obvious, the value returned to 0. P2 denotes energy policy group 2. The energy conservation law promulgated in 1998 worked well, but was amended in 2008. The P2 values were set at 0 before 1998 and at 1 between 1998 and 2006. P3 covers group 3. The medium and long-term specific schema went into operation in 2004. Thus the P3 values are set at 0 before 2004 and 1 from 2004 to 2006. Table 5.45 shows the values of *EI*, real GDP, energy prices, and energy policies.

5.5.3 Energy-Saving Effects of Energy Policies

The effects of energy-saving policies were researched by two methods: (1) with and without antitheses and (2) linear regression.

5.5.3.1 With and without Antitheses

1. Energy policy group 1: The energy intensities from 1991 to 1997 were compared with those from 1982 to 1990 by GM (1,1). Figure 5.5 shows the results. Energy policy reduced energy intensity from 1992 to 1997. From 1991 to 1997, energy-saving effect averaged 0.115 and continues to grow.
2. Energy policy group 2: The energy intensities from 1998 to 2002 were compared with those from 1991 to 1997 by GM (1,1). Figure 5.6 shows the results. From 1998 to 2002, energy-saving effect averaged 0.075, and is decreasing.
3. Energy policy group 3: The energy intensities of 2005 and 2006 were compared with those from 2001 to 2004 by GM (1,1). Figure 5.7 shows the results. The energy-saving effect for 2005 was 0.058, the effect for 2006 was 0.137; the average was 0.097.

We concluded that the energy policies decreased energy intensity. Policy group 1 showed the largest energy saving. However, the method made the important

Table 5.45 *EI*, Real GDP, Energy Price, and Energy Policy, 1982–2006

Year	EI	Real GDP	Energy Policy	P1	P2	P3
1982	2.963	21141.14	1.005	0.0	0.0	0.0
1983	2.818	23435.58	1.068	0.0	0.0	0.0
1984	2.627	26992.17	1.197	0.0	0.0	0.0
1985	2.504	30626.99	1.283	0.0	0.0	0.0
1986	2.425	33336.37	1.342	0.0	0.0	0.0
1987	2.329	37197.81	1.395	0.0	0.0	0.0
1988	2.247	41393.94	1.490	0.0	0.0	0.0
1989	2.250	43075.86	1.615	0.0	0.0	0.0
1990	2.207	44729.55	1.730	0.0	0.0	0.0
1991	2.125	48835.25	2.055	1.0	0.0	0.0
1992	1.957	55789.73	2.370	1.0	0.0	0.0
1993	1.824	63580.38	4.059	1.0	0.0	0.0
1994	1.707	71897.13	6.036	1.0	0.0	0.0
1995	1.645	79751.88	7.316	1.0	0.0	0.0
1996	1.584	87733.86	7.653	1.0	0.0	0.0
1997	1.437	95890.51	8.219	1.0	0.0	0.0
1998	1.279	103401.95	7.644	0.0	1.0	0.0
1999	1.203	111281.00	8.377	0.0	1.0	0.0
2000	1.148	120663.42	12.089	0.0	1.0	0.0
2001	1.096	130678.86	11.980	0.0	1.0	0.0
2002	1.065	142547.21	11.404	0.0	1.0	0.0
2003	1.116	156838.10	13.184	0.0	1.0	0.0
2004	1.177	172655.29	15.056	0.0	1.0	1.0
2005	1.178	190667.99	18.428	0.0	1.0	1.0
2006	1.163	211808.05	22.170	0.0	1.0	1.0

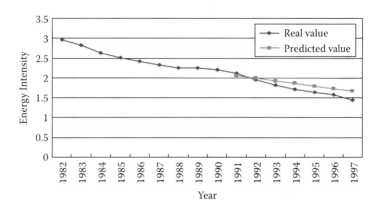

Figure 5.5   Energy-saving effects of energy policy group 1.

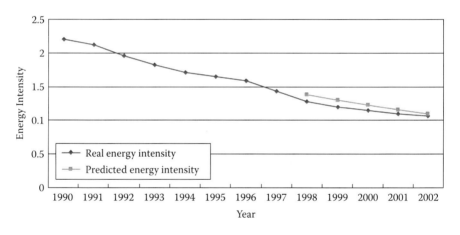

Figure 5.6   Energy-saving effects of energy policy group 2.

assumption that all factors other than the policies were stable. Under real conditions, some of the factors exhibited unstable effects so some errors were generated.

5.5.3.2 Linear Regression

The linear regression model shown below and resolved by SPSS was used to analyze the energy-saving effects. The results of coefficient and related tests are shown in Tables 5.46 and 5.47.

$$EI = 16.312 + 0.00001048 \ GDP - 0.027 \ EP - 0.007 \ TP$$

$$- 0.018 \ SP - 0.121 \ TFP - 0.185 \ P_1 - 0.397 \ P_2 + 0.133 \ P_3$$

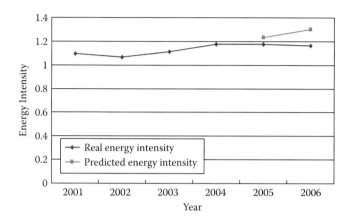

Figure 5.7   Energy-saving effects of energy policy group 3.

Table 5.46 R^2 Test and F Test

			Change Statistic			
R	*R^2*	*Adjusted R^2*	*F Change*	*df1*	*df2*	*Significant F Change*
0.997	0.993	0.990	289.563	8	16	0.000

Table 5.47   Coefficients and *t*-Test

	Unstandardized Coefficient		Standardized Coefficient		
Model	*B*	*Standard Error*	*Beta*	*t*	*Significance*
Constant	16.312	1.675	–	9.739	0.000
GDP	1.048E–5	0.000	0.964	3.842	0.001
EP	−0.027	0.019	−0.270	−1.416	0.176
P1	−0.185	0.061	−0.139	−3.014	0.008
P2	−0.397	0.093	−0.320	−4.272	0.001
P3	0.133	0.083	0.061	1.606	0.128
TFP	−0.121	0.023	−1.320	−5.269	0.000
TP	−0.007	0.017	−0.061	−0.401	0.694
SP	−0.018	0.019	−0.058	−0.935	0.364

Note: Dependent Variable: *EI*.

The adjusted R^2 of 0.993 and the significant F change of 0.000 show that the regression model is effective and the factors explain energy intensity very well. The t-tests show that GDP, TFP, P1, and P2 had significant effects on energy intensity. The effects of EP, TP, SP, and P3 were not significant.

The GDP coefficient is positive. Thus, when other variables are controlled, energy intensity will increase with economic growth. Economic development in China in the past 20 years depended to an extent on large numbers of small-scale factories such as power plants and cement and chemical manufacturing plants. Based on output, these factories consume more energy than large-scale operations. Economic growth means more energy consumption without achieving economies of scale.

The TFP coefficient is negative; when other variables are controlled, energy intensity will decrease as a result of technological progress. Application of new technologies and techniques and the emergence of new products can effectively reduce energy consumption and technological progress will effectively reduce energy consumption.

The coefficients of P1 and P2 are negative. When other variables were controlled, P1 and P2 reduced energy intensity and aided energy saving.

EP had a negative coefficient. When other variables were controlled, energy intensity decreased as energy prices increased. The reason is that price elasticity of energy is negative: as the price of energy increases, consumption will decrease.

The coefficient of TP is negative. When other variables are controlled, energy intensity will decrease with TP increases because tertiary industry consumes less energy than secondary industry. If the proportion of tertiary industry increases, energy consumption will decrease.

The coefficient of SP is negative. When other variables are controlled, energy intensity will decrease with an SP increase. Despite the higher energy consumption by secondary industry, it declines comparatively quickly. Therefore, improving the proportion of secondary industry will reduce the energy intensity.

The P3 coefficient is positive. When other variables are controlled, P3 will increase EI because of the energy intensity increase that occurred after implementation of P3. As shown in Figure 5.4, the growth rate of energy intensity decreased obviously after P3 implementation, so P3 is effective. However, the full effect cannot be determined because P3 was in operation for only 2 years.

5.6 Conclusion

We introduced two methods to evaluate the energy-saving effects from energy polices. The results calculated by different methods were different. The important assumption that energy intensity would be stably affected by other factors may cause the result calculated with and without antitheses to include effects induced by the other factors. The energy-saving effects of P3 are not significant because P3 has been in operation for only 2 years—not long enough to reveal true energy-saving effects.

Calculation with and without antitheses was used to evaluate the short-term effects of energy policy. This method is reasonable if the effects of the other factors are stable over a short period. Linear regression was used to evaluate the long-term effects of energy policies, because it can discern the effects of all factors effectively.

Chapter 6

International Cooperation Project Selection

6.1 Overview

The rapid developments in science and technology, tough international competition, and international production increases have fostered international cooperation. Cooperation in key technology and other industrial fields continues to increase. However, the cooperation in different technology fields shows different characteristics. The cooperation in the domains of general mature physical and chemical technologies, patents, know-how, and general soft technology is generally supported and unlimited, but international cooperation and transfer in the key technology fields are strictly limited. The key technologies are strategic assets and they are under strict government control. The U.S.-led Western developed countries implemented several export control policies covering key technical cooperation with China and the controls affect major scientific and technological planning in China. These controls act as major obstacle to China's technology imports. This chapter focuses on key technology selections.

International science and technology cooperation selection affects every country. Basic research analyzing demands for leading regional industries and dealing with selection of high quality technology and project sources from overseas is minimal. Systematic theories and methods for determining key technology industries through international cooperation are lacking. Studying regional international cooperation project selection based on the international situation and domestic view is crucial.

Regional cooperation project selection methods consist of two phases: (1) international cooperation technological project solicitation and current situation survey and (2) a key technology evaluation expert survey. We can use grey systems and other soft technology approaches to determine key technology areas and priorities to obtain a comprehensive ranking to determine important overseas partner institutions, key support, appropriate technologies, export issues, and partner countries. Research should also lead to relevant policy recommendations.

This chapter discusses foreign key technology selection related to leading industry needs in Jiangsu Province. Jiangsu occupies a leading economic develop position in China. Its rapid development exerted beneficial effects on other provinces and on national economic development. Therefore, the study of regional international technology cooperation project selection based on Jiangsu is relevant.

The Jiangsu Provincial Government's *Eleventh Five Year Technology Development Plan* recommends that Jiangsu's leading industries need focused foreign key technologies and recommends a project team to handle two phases of research work. The first phase focuses mainly on international cooperation technological project solicitation and a current situation survey: We want to first assess the current demand and supply situation of international cooperation in technology and analyze absorption and innovation of international science and technology cooperation and technology transfer in Jiangsu. All these factors play a role in helping enterprises carry out technology transfer and better achieve technical cooperation and absorption and innovation more effectively. The second phase requires a key technology evaluation survey by experts based on the urgency and possibility indices established during the first phase (Figure 6.1).

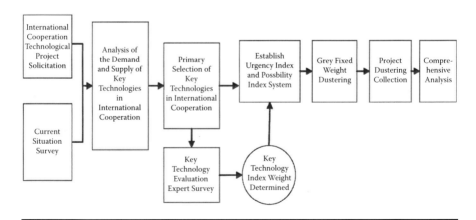

Figure 6.1 Selecting international cooperative key technologies for Jiangsu's leading industries.

6.2 Demand for and Supply of International Cooperative Key Technology in Jiangsu Province

6.2.1 Analysis of Demand

Economic globalization, the rapid development of the socialist market economic system, and waves of regional economic integration are on the rise. The expansion of the regional economies of the Yangtze River Delta, the Pearl River Delta, and Bohai Bay are important sectors of economic development in China. In some sense, the rapid emergence of regional economic cooperation is one result of China's economic and social development. The continuous extension of regional economic cooperation is an effective way to achieve sound and rapid economic development.

In recent years, the international cooperation in Jiangsu is closely linked with the economic and technological strategies. The implementation of a group of international scientific and technological cooperation projects in Jiangsu is very important to ensure better participation in global scientific and technological cooperation and competition and enhance international influence. Cooperation efforts related to international technology and science have greatly improved. Based on a mandated deadline of April 2002, China instituted cooperation measures and exchanges of science and technology with 152 countries and regions; 96 of them signed technological cooperation agreements with China. International scientific and technological cooperation provides technology and techniques, attracts talent and capital, fosters training, and increases exports. It also generates economic and social benefits from new products and technologies, enhances quality, and adjusts the industrial structure. Therefore, regional cooperation in international science and technology is key to regional development.

Jiangsu is a populous and resource-poor province despite long-term rapid economic growth and social and economic development. Various types of conflicts and resource constraints inhibit provincial economic development. Accelerating the strategic adjustment of the industrial structure through international cooperation is a far more important issue. To understand the status of international science and technology cooperation and the current technological demand and supply situations in Jiangsu, the research team conducted a province-wide survey of international technology cooperation status and solicitation efforts.

The survey and data collection targeted enterprises, research institutes, and universities. Jiangsu Province is attentive to the need for international science and technology cooperation; it helps enterprises use the global scientific and technological resources and enhances independent innovation. The government set the state for this new pattern. The enterprises, universities, research institutions, and stakeholders participate in cooperation efforts. The technology solicitation covered seven areas of the province's leading industries:

1. Electronic information: Software, integrated circuits, modern communications, and digital audiovisual technologies and products
2. Modern equipment manufacturing: Computer numerical control (CNC) machining, rail transportation, medical devices, electronic information processing equipment, and instrumentation
3. New materials: Electronic information, high performance metals, chemicals, and medical materials
4. Biotechnology and innovative medicines: New medicines, pesticides, veterinary products, industrial biotechnology
5. New energy sources and energy-saving technology: Solar energy, semiconductor lighting technology, wind and biomass energy, energy-saving and environmental technologies
6. Modern agriculture: Developing new crop varieties, livestock, aquatic farms, Liang Fan technology, agricultural product storage and processing, agricultural equipment, and development of technology
7. Agricultural information technology and application: Post control, control of animal diseases, research and development to find key common technologies
8. Technological innovation in social development: Environment, population and health, economy, living environment, public safety

Through systematic analysis, the team determined the current status of international cooperation, impacts of technology import policies on technological innovation, and current needs for and supplies available for international cooperation. Supply issues will be detailed in the next section. Research and analysis indicate the following needs for Jiangsu Province:

1. Enterprises showed the strongest demands for international science and technology cooperation—more than 50% of the total. Universities and institutes of technology required less. Government organizations such as the Environmental Protection Bureau and Land Resources Bureau) also have demands for international cooperation in science and technology.
2. Demands for equipment manufacturing, modern agriculture, new materials, and technology are relatively strong; demands in the social development domain are relatively weak.
3. Needs for electronic information, new materials, new energy sources and energy-saving technology and equipment arise mainly from enterprises.

Business accounts for more than 60% of the demand for international cooperation in technology. Equipment manufacturing industry accounts for about 92% of international cooperation in technology. Social development, research institutes and universities represent major demands. The research institute demand comprised about 45%, followed by colleges and universities; enterprises accounted for only 11%. Needs for international cooperation in the field of modern agricultural technology mainly came from research institutes (about 65%), followed by enterprises. Universities need a minimum of international cooperation in technology—only 5%. Needs of biotechnology and pharmaceutical innovation and technology in international cooperation in the field are almost the same; needs for companies are slightly higher, about 46%. Universities accounted for about 24% (Figure 6.2).

6.2.2 Analysis of Supply

When economic reform started, international scientific and technological cooperation was limited to government staff. Today, China is at a new stage of carrying out collaborative research projects domestically and overseas and the result has been extensive industrial research.

In Jiangsu Province before 1978, the science and technology cooperation mainly involved Third World and some Eastern European countries. After 1978, it began to vigorously pursue cooperation and exchanges with Western countries. To date it has established scientific and technological cooperation and exchange relations with more than 70 countries and regions. In 2007, Jiangsu Province organized a number of major international science and technology cooperation and exchange activities with Russia, European Union countries, and other key entities. For example, the Jiangsu Province–Novosibirsk Technology Trade Fair was organized jointly with Russia. The Zhenjiang, China Technology

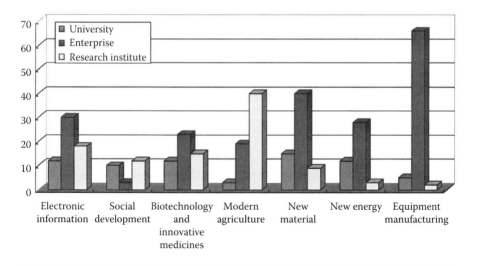

Figure 6.2 Demand distribution of international cooperative technology.

Cooperation Fair was a joint effort of Jiangsu and the European Union. The 2007 Forum on China–EU International Research Cooperation, and European Union's Seventh Research Framework Program involved dissemination of project application information. The Finnish National Technology Agency and the North East of England Development Agency signed a memorandum of comprehensive cooperation has carried out a number of projects. UPM Changshu established research and development (R&D) operations to study nonwood fiber and paper technology through industrial cooperation. The Suzhou Nano Project joined the China–Finland nanotechnology strategic cooperation program. The National Technical Research Centre of Finland (VVT) consulted on the introduction of nanocomposite supersonic thermal spray technology; Dragon Magnet Wire Co. and Taicang are now cooperating. Southeast University and the University of Finland are cooperating on nanotechnology research and application projects. A subsidiary energy company of the University of Newcastle in the United Kingdom and Nantong University signed an agreement to build a Sino-British Jiangsu New Energy Research Center.

Jiangsu Province has also actively encouraged the establishment of foreign, cooperative, and independent R&D institutions. Currently, more than 100 foreign R&D institutions, mainly from the United States, Japan, Europe, Taiwan, and Hong Kong currently operate in Jiangsu. R&D activities focus on electronic information, advanced manufacturing and automation, medicines, and other industries. The world's top 500 foreign enterprises account for 17%, foreign-owned R&D institutions constitute 65%, and foreign R&D represents the balance. Jiangsu's cooperation activities achieved considerable promotion, demonstration, and encouragement of scientific and technological cooperation. Active exchanges among research institutions, academic organizations, enterprises, cities, and individual scientists continue. In this comprehensive, multichannel, multiform, and multilevel international scientific and technological cooperation, Jiangsu's economic and social development has produced significant results. The supply situation can be summarized as follows:

1. International cooperation in science and technology is mainly with overseas enterprises; cooperation arrangements with overseas universities are fewer.
2. The main international cooperation countries are the United States, Germany, Japan, Russia, Britain, Canada, Korea, France, Netherlands, Italy, Hong Kong, Australia, Israel, Taiwan, Switzerland, and Sweden. These countries account for 90% of partner activities. The United States is the major partner country. Cooperation projects with Vietnam, Thailand, Cuba, and other Third World countries involve the introduction of modern agriculture or technology outputs.
3. Overseas partner institutions are in 20 countries: Austria, Australia, Commonwealth of Independent States (Belarus, Russia, and Ukraine), Northern Europe (Denmark, Norway, and Sweden), Germany, France,

Korea, Canada, Czech Republic United States, Japan, Switzerland, Israel, Italy, Britain, and Taiwan.

4. Overseas partner institutions are mainly involved in equipment manufacturing and new materials, followed by distribution in the new energy, electronic information, biological and pharmaceutical fields, modern agriculture. Few overseas institutions Social development represents a low percentage of overseas institutions.

6.2.3 Selection of Key Technologies for International Cooperation by Leading Regional Industries

In the initial screening of the collected technology projects, the ability of independent innovation should hold a prominent position based on required inputs and outputs, localization issues, and other parameters of projects considered. We should follow normative and effective principles when screening technology and determine potential impacts on regional economic development. In this regard, this book will utilize Professor Liu Sifeng's leading industry supply selection of key technologies to study the primary principles of the international cooperation key technologies for leading industries for the region.

What is the status of international cooperation with the current demand for technology and what will the status be in the future for Jiangsu Province? Professor Liu's research group carried out an international cooperation technological project solicitation and current situation survey in Jiangsu Province targeting enterprises, research institutes, and universities. The scope of collection covered the seven leading industrial areas in Jiangsu: electronic information, modern equipment manufacturing, new materials, biotechnology and innovative drugs, new energy and energy saving technology, modern agriculture and social development (Figure 6.3).

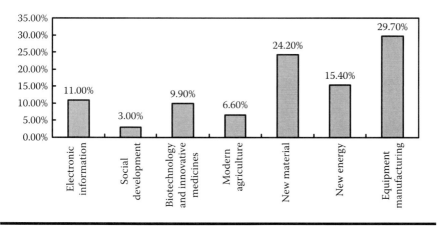

Figure 6.3   Area distribution of overseas partner institutions.

A total of 327 valid questionnaires were analyzed. Of a total of 371 effective techniques, 308 were class A technologies (a good foundation for international cooperation) and 63 were class B technologies (needed for the province's leading industries but not useful for international cooperation). The team developed fundamental principles. If a technology listed one of the following conditions, it was disqualified and not submitted to experts for evaluation:

1. The capacity for secondary development and innovation capacity was weak and direct use of production line technology was emphasized or the intent of the cooperative arrangement was for the partner country to sell a mature technology patent or advanced equipment with no opportunity for innovation.
2. The key technology was already mature and international cooperation was not needed.
3. The proposal involved pure theory that would be difficult to advance to manufacturing level within 3–5 years.
4. The technology aims at solving specific problems not related to leading industry key technologies appropriate Jiangsu Province.
5. The technology is established and successfully produced.
6. The technology does not meet the requirements for production or lacks a sufficient technical basis. B class technologies must contribute to high-end development in Jiangsu in the near future.
7. An agricultural technology project that involves only acquisition of seed sources from partner countries and includes no plans for technical cooperation aimed at improving plant varieties. Agricultural technology in Jiangsu province requires breeding of plant varieties suitable for specific soil and climate conditions.

6.3 Demand and Supply Indices of International Cooperative Key Technology

6.3.1 The Principle of Index System Construction

The scientific basis and rationality of an index system directly affect the scientific validity of an evaluation and the achievement of the research objective. The basic factors that deserve consideration when choosing regional international collaborative key technology include: propelling development of regional economic technology and establishing new industries, aiding environmental protection and protecting the population's health; obtaining overseas technology resources, reliable funding, management, and work force, and a product or service that could be on the market in 3 to 5 years. The principles required to establish urgency and possibility indices are:

6.3.1.1 Integrity

The evaluation index system should cover all aspects of the key technology. An index should be representative, logically organized, and relate to other indices to produce a clear picture of a technology.

6.3.1.2 Scientific Rationality

Scientific rationality is an important aspect of credibility. A system for evaluating an index system of regional international cooperative key technology should be based on scientific theory, the basic characteristic of the key technology, the present status and capabilities of dominating industries, correlative research achievement overseas, and the opinions of specialists in the field. Potential scheduling, supply issues, and other factors that may interfere with the index system should be adjusted to reduce errors and ensure scientific rationality.

6.3.1.3 Index System

The factors that affect and determine the performance of international cooperation arrangements are comprehensive. Thus, statistical design and systematic evaluation are required to ensure a rational arrangement of an index system to accurately reflect the supply-and-demand status of a project and provide essential data support.

6.3.1.4 Independence

The index system should reflect urgency and consist of a series of related but independent individual indices that are clear and relevant.

6.3.1.5 Feasibility

Feasibility requires easy operation. An index may be appropriate but lack operability; it will not effectively evaluate parameters. Data in statistical annals should be considered.

6.3.1.6 Comparability

The choice of a regional international cooperative key technology is based to an extent on comparisons. To build an index system to objectively reflect the various aspects of a technology, the evaluating indices should be comparable with each other and among projects.

6.3.2 Evaluating Indices of Urgency and Possibility

To objectively quantify the possibilities for acquiring regional international cooperative key technology, we applied Delphi methodology to establish an international cooperative key technology index system on the basis of supply-and-demand status of key technologies in Jiangsu province. The index system is shown in Figure 6.4.

According to five evaluation criteria—significance, universality, propelling ability, practicality, and social value—the urgency of regional cooperative key technology was evaluated by the indices listed in Table 6.1.

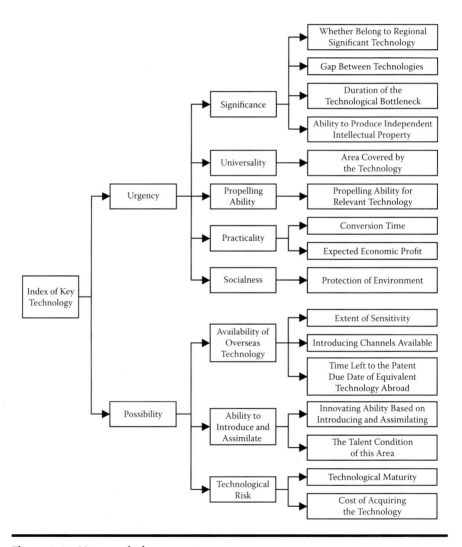

Figure 6.4   Urgency index system.

Table 6.1   Urgency Index System

Criterion	Index	Evaluation
Significance	Belonging to regional significant technology	A = yes; B = no
	Gap between technologies	A = more than 10 years; B = 5 to 10 years; C = 3 to 5 years; D = less than 3 years
	Duration of technological bottleneck	A = more than 10 years; B = 5 to 10 years; C = 3 to 5 years; D = less than 3 years
	Ability to produce independent intellectual property	A = intellectual property may be completely owned; B = intellectual property may be partially owned; c = intellectual property may not be owned
Universality	Area covered by technology	A = generally used in many areas; B = generally used in industry; C = special technology
Propelling Ability	Propelling relevant technology	A = very strong; B = strong; C = common; D = weak
Practice	Conversion time	A = less than 1 year; B = 1 to 3 years; C = 4 to 5 years; D = more than 5 years
	Expected profit (RMB)	A = more than 5 million; B = 3 to 5 million; C = 1 to 3 million; D = less than 1 million
Social value	Protection of environment	A = very strong; B = strong; C = common; D = weak

The urgency indices are as follows:

1. Belonging to regional significant technology: whether the technology is suitable.
2. Gap between technologies: time between acquisition of technology and its development in Jiangsu.
3. Duration of technological bottleneck: time between development of technological bottleneck and obstruction of relevant industries.
4. Ability to produce independent intellectual property: potential for ownership of independent intellectual property.

5. Area covered by technology: potential for use in related fields.
6. Propelling ability: ability of technology to propel upstream and downstream technologies.
7. Conversion time: time needed for assimilation and industrialization of technology.
8. Expected profit: yield from technology. Annual output value should be calculated if the technology can be industrialized; reduced cost should be calculated if the technology cannot be industrialized.
9. Protection of environment: whether water, gas, noise and radiation contamination will be reduced after project is implemented.

Table 6.2 evaluates regional cooperative key technology possibilities based on the availability of overseas technology, ability to introduce and assimilate, and technological risk. The possibility indices are:

1. Extent of sensitivity: sensitivity to community issues and customs.
2. Introducing channels available: methods for introducing technology.

Table 6.2   Possibility Index

Criterion	Index	Evaluation
Availability of overseas technology	Extent of sensitivity	A = nonsensitive; B = commonly sensitive; C = quite sensitive; D = very sensitive
	Introducing methods available	A = several channels; B = single channel; C = no channels
	Time left to patent due date of equivalent technology abroad	A = less than 3 years; B = 3 to 5 years; C = 5 to 10 years; D = more than 10 years
Ability to Introduce and assimilate	Innovating ability based on introducing and assimilating	A = very strong; B = strong; C = common; D = weak
	Talent in technological area	A = very good; B = good; C = poor
Technological risk	Technological maturity	A = very mature; B = mature; C = not mature
	Cost of acquiring technology (RMB)	A = less than 1 million; B = 1 to 3 million; C = 3 to 5 million; D = more than 5 million

3. Time left to patent due date of equivalent technology abroad: duration of use of technology.
4. Innovating ability based on introducing and assimilating: using technology as a basis for innovation and industrialization.
5. Talent condition of this technological area: talent required to convert technology.
6. Technological maturity: worldwide technological maturity.
7. Cost of acquiring: direct cost of acquiring technology or cost of cooperative R&D.

6.3.3 Weights of Urgency and Possibility Indices

6.3.3.1 Determining Weights

After establishing an index system to assess international cooperative key technology, the weight of urgency and possibility indices should be determined. The first step is distribution of questionnaires to specialists. Each questionnaire lists indices and provides blank boxes for experts to use to assign weights. Table 6.3 shows the urgency index weights and Table 6.4 shows the urgency index weights.

After the questions are completed, the next step is statistical analysis to calculate mean values (index weights) and variances.

Example 6.1

The following table shows the index weights assigned by 10 specialists.

Specialist	1	2	3	4	5	6	7	8	9	10
Index Weight	0.1	0.13	0.095	0.12	0.1	0.11	0.12	0.09	0.13	0.14

Solution

The index weight is

$$(0.1 \times 2 + 0.13 \times 2 + 0.095 + 0.12 \times 2 + 0.11 + 0.09 + 0.14)/10 = 0.1135$$

The weight distribution graph indicates that the means and variances reveal a normal distribution, fixed weights of certain indices that deviated distinctly, and finally the weights of urgency and possibility indices.

6.3.3.2 Analysis of Urgency and Possibility Index Weights

The weights of the urgency and possibility indices allowed us to quantify the choices of key technologies and eventually make a final choice. Table 6.5 shows the urgency weights for dominant regional industries.

Table 6.3   Weights of Urgency Indices

Index	Belonging to regional significant technology	Gap between technologies	Duration of technological bottleneck	Ability to produce independent intellectual property	Area covered by technology
Weight					
Index	Ability to propel relevant technology	Conversion time	Expected economic profit	Protection of environment	
Weight					

Table 6.4   Weights of Possibility Indices

Index	Extent of sensitivity	Introducing methods available	Time left to patent due date of equivalent technology abroad	Innovating ability based on introducing and assimilating
Weight				
Index	Talent available in area	Technological maturity	Technology acquisition cost	
Weight				

The resulting index weights for possibilities for regional industries are shown in Table 6.6.

6.3.3.3 Relative Importance of Urgency and Possibility Indices

The relative importance of both indices was surveyed by a questionnaire. The results are shown in Table 6.7.

The initial relative importance was set at 1 and the specialists used their judgment to complete the table. The distribution of the sample is shown in Figure 6.5. The results indicate that the equivalence of relative importance of the urgency index is 1, and the equivalence for the possibility index is 1.23.

Table 6.5 Weights of Urgency Indices for Dominant Regional Industries

Index	Belonging to regional significant technology	Gap between technologies	Duration of technological bottleneck	Ability to produce independent intellectual property	Area covered by technology
Weight	0.15	0.1	0.09	0.14	0.09
Index	Ability to propel relevant technology	Conversion time	Expected profit	Protection of environment	
Weight	0.11	0.07	0.13	0.12	

Table 6.6 Possibility Index Weights

Index	Extent of sensitivity	Introducing methods available	Time left to patent due date of equivalent technology abroad	Innovating ability based on introducing and assimilating
Weight	0.14	0.10	0.11	0.18
Index	Talent condition of area	Technological maturity	Cost of acquiring technology	
Weight	0.18	0.17	0.12	

Table 6.7 Comparison of Urgency and Possibility

Index	Urgency Index	Possibility Index
Relative importance	1	

6.4 Choosing Model Based on Grey Clustering with Fixed Weight

The effective choice of a key technology requires an extensive index system containing several indices of different magnitudes. Grey fixed weight clustering is useful for solving problems of multiple indices of different magnitudes via effective classification of observed objects.

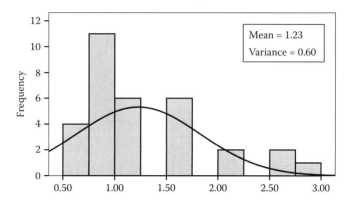

Figure 6.5 **Sample distribution of relative importance of urgency and possibility indices.**

6.4.1 Methodology

The main idea of grey clustering with fixed weight is to treat n preliminary selected technical projects as clustering objects. The m clustering indices are based on the evaluating index system; s denotes grey classes of objects i into class k ($k \in \{1,2,\cdots,s\}$) according to the observation x_{ij} ($i = 1,2,\cdots,n$; $j = 1,2\cdots,m$) of object i about index j ($j = 1,2,\cdots,m$). The ordering of key technologies of regional international cooperation is achieved by analyzing the clustering coefficients of different grey classifications. The key technology choosing algorithm based on grey clustering with fixed weight is:

Step 1: Divide classification s according to evaluation requirements.
Step 2: Assign a whitenization weight function of each index according to its value range. The whitenization weight function of subclass k of index j generally falls into one of four types:
(i) The typical whitenization weight function (Figure 6.6) is shortened as $f_j^k(\cdot)(j=1,2,\cdots,m;k=1,2\cdots,s)$, that is,

$$f_j^k(x) = \begin{cases} 0 & x \notin \left[x_j^k(1), x_j^k(4)\right] \\[2ex] \dfrac{x - x_j^k(1)}{x_j^k(2) - x_j^k(1)} & x \in \left(x_j^k(1), x_j^k(2)\right) \\[2ex] 1 & x \in \left[x_j^k(2), x_j^k(3)\right] \\[2ex] \dfrac{x_j^k(4) - x}{x_j^k(4) - x_j^k(4)} & x \in \left(x_j^k(3), x_j^k(4)\right) \end{cases} \tag{6.1}$$

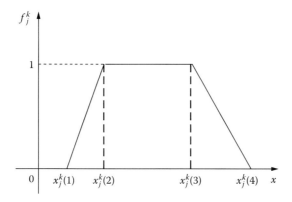

Figure 6.6 Typical whitenization weight function.

(ii) Lower measure whitenization weight function (Figure 6.7) is shortened as $\left[-,-,x_j^k(3),x_j^k(4)\right]$, that is,

$$
f_j^k(x)=\begin{cases} 0 & x\notin\left[1,x_j^k(4)\right] \\[2mm] 1 & x\in\left[1,x_j^k(3)\right] \\[2mm] \dfrac{x_j^k(4)-x}{x_j^k(4)-x_j^k(3)} & x\in\left(x_j^k(3),x_j^k(3)\right] \end{cases} \tag{6.2}
$$

(iii) Moderate measure whitenization weight function (Figure 6.8) is shortened as $\left[x_j^k(1),x_j^k(2),-x_j^k(4)\right]$, that is,

$$
f_j^k(x)=\begin{cases} 0 & x\notin\left[x_j^k(1),x_j^k(4)\right] \\[2mm] \dfrac{x-x_j^k(1)}{x_j^k(2)-x_j^k(1)} & x\in\left[x_j^k(1),x_j^k(2)\right] \\[2mm] \dfrac{x_j^k(4)-x}{x_j^k(4)-x_j^k(2)} & x\in\left[x_j^k(2),x_j^k(4)\right] \end{cases} \tag{6.3}
$$

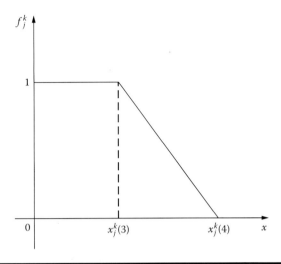

Figure 6.7   Lower measure whitenization weight function.

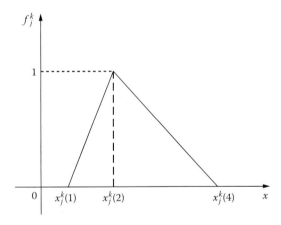

Figure 6.8   Moderate whitenization weight function.

(iv) Upper measure whitenization weight function (Figure 6.9) is short-ened as $\left[x_j^k(1), x_j^k(2), -, - \right]$, that is,

$$
f_j^k(x) = \begin{cases} 0 & ,x \notin \left[x_j^k(1), 7 \right] \\[2ex] \dfrac{x - x_j^k(1)}{x_j^k(2) - x_j^k(1)} & ,x \in \left[x_j^k(1), x_j^k(2) \right] \\[2ex] 1 & ,x \in \left(x_j^k(2), 7 \right] \end{cases} \tag{6.4}
$$

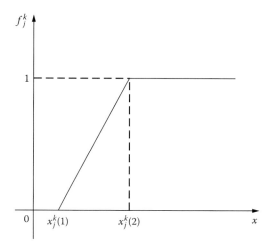

Figure 6.9   Upper measure whitenization weight function.

Step 3: Determine the clustering weight $\eta_j(j = 1,2\ldots,m)$ of each index.

Step 4: Based on the whitenization weight function $f_j^k(\cdot)(j = 1,2,\cdots,m; k = 1,2,\cdots,s)$ determined from Steps 1 and 2, the clustering weight $\eta_j(j = 1,2\ldots,m)$ and observation $x_{ij}(i = 1,2,\cdots,n; j = 1,2,\cdots,m)$ of object i about index $j(j = 1,2,\cdots,m)$ and calculate the grey fixed weight clustering coefficient

$$\sigma_i^k = \sum_{j=1}^{m} f_j^k(x_{ij})\eta_j, (i = 1,2,\cdots,n; k = 1,2,\cdots s)$$

Step 5: Calculate the clustering weight vector according to the fixed weight coefficient of each class

$$\sigma_i = (\sigma_i^1, \sigma_i^2, \cdots, \sigma_i^s) = \left(\sum_{j=1}^{m} f_j^1(x_{ij})\eta_j^1, \sum_{j=1}^{m} f_j^2(x_{ij})\eta_j^2, \cdots, \sum_{j=1}^{m} f_j^s(x_{ij})\eta_j^s \right)$$

Step 6: Calculate the clustering coefficient matrix

$$\sum = (\sigma_i^k) = \begin{bmatrix} \sigma_1^1 & \sigma_1^2 & \cdots & \sigma_1^s \\ \sigma_2^1 & \sigma_2^2 & \cdots & \sigma_2^s \\ \vdots & \vdots & \ddots & \vdots \\ \sigma_n^1 & \sigma_n^2 & \cdots & \sigma_n^s \end{bmatrix}$$

Step 7: Determine the class to which objects belong based on the clustering coefficient matrix. Object i belongs to class k^* if $\max_{1 \leq k \leq s}\{\sigma_i^k\} = \sigma_i^{k^*}$.

Step 8: Determine the priority orders of objects based on class and clustering coefficient value.

6.4.2 Priority Orders of Key Technology

The ordering of international cooperative key technology is based on a large index containing several indices with different dimensions. Grey clustering may weaken the influences of some indices that may be handled via grey clustering with fixed weight. Determining priority orders by grey clustering with fixed weight follows the following steps (Figure 6.10).

First, the clustering objects should be selected. According to the evaluating index system described in Section 6.2, the dominant industries are electronic information, modern equipment manufacturing, new materials, biotechnology and innovative medicines, new energy sources and energy-saving technology, modern agriculture, and social development. The next step is developing the evaluating index. The urgency indices evaluate whether an industry belongs to regional significant technology, then determines the gap between technologies, duration of technological bottlenecks, ability to produce independent intellectual property, area covered by the technology, its ability to propel relevant technology, conversion time, expected economic profit, and protection of environment. The possibility index covers sensitivity, introducing channels available, time left to patent due

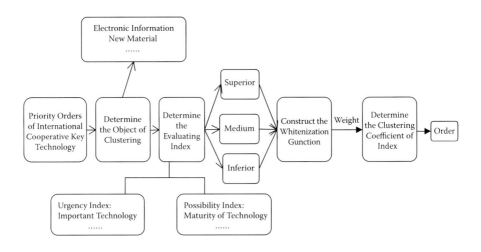

Figure 6.10 Priority orders of key technology based on grey fixed weight clustering.

date of equivalent technology abroad, innovating ability based on introducing and assimilating, talent in the technological area, technological maturity, and cost of acquiring the technology.

To determine the priority orders of key technologies, the index system is divided into three grey classes: inferior, medium, and superior. The lower measure whitenization weight function is used to describe the inferior class; the moderate function is used to describe the medium class; and the upper measure function is used to describe the superior class (Figure 6.11).

The order is based on the specialists' evaluations of the technologies. According to the urgency and possibility indices and questionnaire answers, the relative importance of the urgency index σ and the possibility index ρ1 can be calculated via the concrete whitenization function to yield index weight:

$$\sigma_i^k = \sum_{j=1}^{m} f_j^k(x_{ij})\eta_j$$

According to formula:

$$\theta^j = (\sigma^j * 1 + \rho^j * \mu) / (1 + \mu)$$

the comprehensive index θ may be calculated. According to formula:

$$\max_{1 \le k \le s}\{\sigma_i^k\} = \sigma_i^{k^*}$$

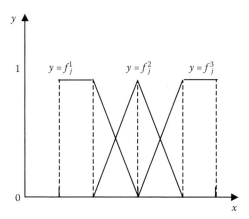

Figure 6.11 Value of whitenization weight function.

We can determine to which grey class urgency index the possibility index and comprehensive index, respectively, belong. Grey fixed weight clustering may be used to calculate the clustering coefficients of the urgency, possibility, and comprehensive indices of each project to yield project sequence.

6.5 Selection of International Cooperative Key Technology

High-speed development of technology and the increasingly intensive competition require international technological cooperation. By inducing advanced technology through international cooperation, technology-importing countries can cultivate quality technical talent. The enterprises that expand their technologies can gain competitive advantages domestically and abroad through innovation. Technology-exporting countries can utilize international technological cooperation to increase domestic job opportunities, prolong technology life cycles, and propel innovations. However, the demands for and availability of key technology differ among regions. Jiangsu Province's selection of key technology for international collaboration is an example.

In Jiangsu, 50 projects in 7 fields were selected as research objects. Using Section 5.2 as a reference, nine indices were chosen to evaluate urgency: (1) whether belongs to regional significant technology; (2) gap between technologies; (3) duration of technological bottleneck; (4) ability to produce independent intellectual property; (5) area covered by technology; (6) ability to propel relevant technology; (7) conversion time; (8) expected economic profit; and (9) protection of environment. The seven indices selected to evaluate possibilities were: (1) extent of sensitivity; (2) introducing channels available; (3) time left to patent due date of equivalent technology abroad; (4) innovating ability based on introducing and assimilating; (5) talent condition of technological area; (6) technological maturity; and (7) cost of acquiring technology.

The evaluation indices of key technologies were based on expert choices. For indices of significant technology, A is valued at 7 and B is valued at 1. For other indices, A is valued at 7, B is valued at 5, C is valued at 3, and D is valued at 1 (Table 6.8).

The data of grey fixed weight clustering methodology were based on specialists' evaluations of key technologies. Five specialists evaluated project 1 (Tables 6.9 and 6.10). We calculated the comprehensive evaluations of experts based on table data (the

Table 6.8 Values of Indices

	A	B	C	D
Indices of key technologies	7	1	–	–
Other indices	7	5	3	1

Table 6.9 Experts' Urgency Indices for Project 1

Expert	Belonging to Regional Significant Technology	Gap between Technologies	Duration of Technological Bottleneck	Ability to Produce Independent Intellectual Property	Area Covered by Technology
1	B	B	B	C	B
2	B	C	B	B	C
3	B	D	D	B	B
4	B	C	C	B	A
5	A	D	D	B	B

Expert	Ability to Propel Relevant Technology	Conversion Time	Expected Economic Profit	Protection of Environment
1	D	C	B	A
2	C	C	D	A
3	C	B	C	D
4	B	A	A	C
5	C	A	D	A

Table 6.10 Experts' Possibility Indices for Project 1

Expert	Extent of Sensitivity	Introducing Methods Available	Time Left to Patent Due Date of Equivalent Technology Abroad	Innovating Ability Based on Introducing and Assimilating
1	A	A	C	C
2	A	A	A	B
3	A	A	A	B
4	C	A	B	A
5	A	A	A	C
Expert	Available Talent in area	Technological Maturity	Acquisition Cost	
1	C	B	C	
2	B	B	B	
3	A	B	B	
4	A	B	C	
5	B	A	A	

mean value is calculated as the comprehensive evaluation). For instance, for indices belonging to significant technological areas:

$$x_1 = \frac{1}{5}(1+1+1+1+7) = 2.2$$

The comprehensive urgencies of other indices could be calculated in the same way; they are 2.6, 3, 4.6, 5, 3, 5, 3.4, and 5. The comprehensive possibility indices are 6.2, 7, 5.8, 4.6, 5.4, 5.4, and 4.6.

Indices fall into three grey classes: inferior, medium, and superior. The lower measure whitenization weight function is used to describe the inferior class; the moderate measure whitenization weight function is used to describe the medium class; and the upper measure whitenization weight function is used to describe the superior class. According to the design criteria, the grey internal is [1, 7] for building the concrete whitenization function of each grey class based on the distribution of each index. The calculation methods are as follows:

1. Inferior

$$f(x) = \begin{cases} 1, & 1 < x < 2 \\ 2 - 0.5x, & 2 < x < 4 \\ 0, & \text{others} \end{cases}$$

2. Medium

$$f(x) = \begin{cases} 0.5x - 1, & 2 < x < 4 \\ 3 - 0.5x, & 4 < x < 6 \\ 0, & \text{others} \end{cases}$$

3. Superior

$$f(x) = \begin{cases} 0.5x - 2, & 4 < x < 6 \\ 1, & 6 < x < 7 \\ 0, & \text{others} \end{cases}$$

The whitenization functions of other indices are listed in Tables 6.11 and 6.12. According to experts' evaluations, the urgency index weights are

$$w = (0.15, 0.1, 0.09, 0.14, 0.09, 0.11, 0.07, 0.13, 0.12)$$

The possibility index weights are

$$w = (0.14, 0.10, 0.11, 0.18, 0.18, 0.17, 0.12)$$

The ratio of urgency index to possibility index is 1:1.23 based on

$$\sigma_i^k = \sum_{j=1}^{m} f_j^k(x_{ij}) \cdot \eta_j$$

The clustering coefficient of inferior urgency of project 1 is

$$\sigma^1 = 0.9 * 0.15 + 0.4 * 0.1 + 0 * 0.9 + 0.4 * 0.14$$

$$+ 0 * 0.9 + 0.5 * 0.11 + 0 * 0.07 + 0.6 * 0.13 + 0 * 0.12$$

$$= 0.364$$

Table 6.11 Whitenization Functions of Possibility Indices

Index	Inferior	Medium	Superior
Gap between technologies	$f(x) = \begin{cases} 1, & 1 < x < 2 \\ 2 - 0.5x, & 2 < x < 4 \\ 0, & \text{others} \end{cases}$	$f(x) = \begin{cases} 0.5x - 1, & 2 < x < 4 \\ 3 - 0.5x, & 4 < x < 6 \\ 0, & \text{others} \end{cases}$	$f(x) = \begin{cases} 0.5x - 2, & 4 < x < 6 \\ 1, & 6 < x < 7 \\ 0, & \text{others} \end{cases}$
Duration of technological bottleneck	$f(x) = \begin{cases} 1, & 1 < x < 2 \\ 3 - x, & 2 < x < 3 \\ 0, & \text{others} \end{cases}$	$f(x) = \begin{cases} x - 2, & 2 < x < 3 \\ 4 - x, & 3 < x < 4 \\ 0 & \text{others} \end{cases}$	$f(x) = \begin{cases} 0.5x - 1.5, & 3 < x < 5 \\ 1, & 5 < x < 7 \\ 0, & \text{others} \end{cases}$
Ability to produce independent intellectual property	$f(x) = \begin{cases} 1, & 1 < x < 2 \\ 3 - x, & 2 < x < 3 \\ 0, & \text{others} \end{cases}$	$f(x) = \begin{cases} x - 2, & 2 < x < 3 \\ 4 - x, & 3 < x < 4 \\ 0 & \text{others} \end{cases}$	$f(x) = \begin{cases} 0.5x - 1.5, & 3 < x < 5 \\ 1, & 5 < x < 7 \\ 0, & \text{others} \end{cases}$
Area covered bytechnology	$f(x) = \begin{cases} 1, & 3 < x < 4 \\ 5 - x, & 4 < x < 5 \\ 0, & \text{others} \end{cases}$	$f(x) = \begin{cases} x - 4, & 4 < x < 5 \\ 6 - x, & 5 < x < 6 \\ 0 & \text{others} \end{cases}$	$f(x) = \begin{cases} x - 5, & 5 < x < 6 \\ 1, & 6 < x < 7 \\ 0, & \text{others} \end{cases}$

Criterion			
Ability to propel relevant technology	$f(x) = \begin{cases} 1, & 3 < x < 4 \\ 5 - x, & 4 < x < 5 \\ 0, & \text{others} \end{cases}$	$f(x) = \begin{cases} x - 4, & 4 < x < 5 \\ 6 - x, & 5 < x < 6 \\ 0 & \text{others} \end{cases}$	$f(x) = \begin{cases} x - 5, & 5 < x < 6 \\ 1, & 6 < x < 7 \\ 0, & \text{others} \end{cases}$
Conversion time	$f(x) = \begin{cases} 1, & 1 < x < 2 \\ 2 - 0.5x, & 2 < x < 4 \\ 0, & \text{others} \end{cases}$	$f(x) = \begin{cases} 0.5x - 1, & 2 < x < 4 \\ 3 - 0.5x, & 4 < x < 6 \\ 0 & \text{others} \end{cases}$	$f(x) = \begin{cases} 0.5x - 2, & 4 < x < 6 \\ 1, & 6 < x < 7 \\ 0, & \text{others} \end{cases}$
Expected profit	$f(x) = \begin{cases} 1, & 1 < x < 2 \\ 2 - 0.5x, & 2 < x < 4 \\ 0, & \text{others} \end{cases}$	$f(x) = \begin{cases} 0.5x - 1, & 2 < x < 4 \\ 3 - 0.5x, & 4 < x < 6 \\ 0 & \text{others} \end{cases}$	$f(x) = \begin{cases} 0.5x - 2, & 4 < x < 6 \\ 1, & 6 < x < 7 \\ 0, & \text{others} \end{cases}$
Protection of environment	$f(x) = \begin{cases} 1, & 1 < x < 3 \\ 4 - x, & 3 < x < 4 \\ 0, & \text{others} \end{cases}$	$f(x) = \begin{cases} x - 3, & 3 < x < 4 \\ 3 - 0.5x, & 4 < x < 6 \\ 0 & \text{others} \end{cases}$	$f(x) = \begin{cases} 0.5x - 2, & 4 < x < 6 \\ 1, & 6 < x < 7 \\ 0, & \text{others} \end{cases}$

Table 6.12 Whitenization Functions of Possibility Indices

Index	Inferior	Medium	Superior
Extent of Sensitivity	$f(x) = \begin{cases} 1, & 3 < x < 4 \\ 5-x, & 4 < x < 5 \\ 0, & \text{others} \end{cases}$	$f(x) = \begin{cases} x-4, & 4 < x < 5 \\ 3.5-0.5x, & 5 < x < 7 \\ 0 & \text{others} \end{cases}$	$f(x) = \begin{cases} 0.5x-2.5, & 5 < x < 7 \\ 0, & \text{others} \end{cases}$
Introducing Methods Available	$f(x) = \begin{cases} 1, & 1 < x < 5 \\ 6-x, & 5 < x < 6 \\ 0, & \text{others} \end{cases}$	$f(x) = \begin{cases} x-5, & 5 < x < 6 \\ 7-x, & 6 < x < 7 \\ 0 & \text{others} \end{cases}$	$f(x) = \begin{cases} x-6, & 6 < x < 7 \\ 0, & \text{others} \end{cases}$
Time Left to Patent Due Date of Equivalent Technology Abroad	$f(x) = \begin{cases} 1, & 1 < x < 3 \\ 2.5-0.5x, & 3 < x < 5 \\ 0, & \text{others} \end{cases}$	$f(x) = \begin{cases} 0.5x-1.5, & 3 < x < 5 \\ 3.5-0.5x, & 5 < x < 7 \\ 0 & \text{others} \end{cases}$	$f(x) = \begin{cases} 0.5x-2.5, & 5 < x < 7 \\ 0, & \text{others} \end{cases}$
Innovating Ability Based on Introducing and Assimilating	$f(x) = \begin{cases} 1, & 3 < x < 4 \\ 5-x, & 4 < x < 5 \\ 0, & \text{others} \end{cases}$	$f(x) = \begin{cases} x-4, & 4 < x < 5 \\ 6-x, & 5 < x < 6 \\ 0 & \text{others} \end{cases}$	$f(x) = \begin{cases} x-5, & 5 < x < 6 \\ 1, & 6 < x < 7 \\ 0, & \text{others} \end{cases}$

Available talent in area	$f(x) = \begin{cases} 1, & 3 < x < 4 \\ 5-x, & 4 < x < 5 \\ 0, & \text{others} \end{cases}$	$f(x) = \begin{cases} x-4, & 4 < x < 5 \\ 6-x, & 5 < x < 6 \\ 0 & \text{others} \end{cases}$	$f(x) = \begin{cases} x-5, & 5 < x < 6 \\ 1, & 6 < x < 7 \\ 0, & \text{others} \end{cases}$
Technological Maturity	$f(x) = \begin{cases} 1, & 3 < x < 4 \\ 5-x, & 4 < x < 5 \\ 0, & \text{others} \end{cases}$	$f(x) = \begin{cases} x-4, & 4 < x < 5 \\ 6-x, & 5 < x < 6 \\ 0 & \text{others} \end{cases}$	$f(x) = \begin{cases} x-5, & 5 < x < 6 \\ 1, & 6 < x < 7 \\ 0, & \text{others} \end{cases}$
Acquisition cost	$f(x) = \begin{cases} 1, & 1 < x < 3 \\ 2.5-0.5x, & 3 < x < 5 \\ 0, & \text{others} \end{cases}$	$f(x) = \begin{cases} 0.5x-1.5, & 3 < x < 5 \\ 3.5-0.5x, & 5 < x < 7 \\ 0 & \text{others} \end{cases}$	$f(x) = \begin{cases} 0.5x-2.5, & 5 < x < 7 \\ 0 & \text{others} \end{cases}$

The clustering coefficient of inferior possibility of project 1 is

$$\rho^1 = 0*0.14 + 0*0.1 + 0*0.11 + 0.4*0.18 + 0*0.18$$
$$+ 0*0.17 + 0.2*0.12$$
$$= 0.096$$

The clustering coefficient of inferior comprehensiveness of project 1 is

$$\theta^1 = (\sigma^1 * 1 + \rho^1 * 1.23)/(1 + 1.23)$$
$$= (0.364 + 0.096 * 1.23)/(1 + 1.23) = 0.2162$$

Similarly, the clustering coefficient of medium urgency, possibility, and comprehensive class of project 1 is:

$$\sigma^2 = 0.601, \quad \rho^2 = 0.536, \quad \theta^2 = 0.5651$$

The clustering coefficient of superior urgency, possibility, and comprehensiveness class of project 1 is:

$$\sigma^3 = 0.035, \quad \rho^3 = 0.368, \quad \theta^3 = 0.2187$$

Thus the clustering coefficient vectors of urgency, possibility, and comprehensiveness index are:

$$\sigma = (\sigma^1, \sigma^2, \sigma^3) = (0.364, 0.601, 0.035)$$

$$\rho = (\rho^1, \rho^2, \rho^3) = (0.096, 0.536, 0.368)$$

$$\theta = (\theta^1, \theta^2, \theta^3) = (0.216, 0.565, 0.219)$$

The evaluations of other key technologies were conducted in the same way. The 50 projects were ordered by their comprehensiveness indices. The results appear in Tables 6.13 through 6.19.

The clustering coefficients of urgency, possibility, and comprehensive indices are shown in Table 6.20. According to

Table 6.13 Comprehensive Evaluation of Key Technology in Electronic Information Field

		Project 1	Project 2	Project 3	Project 4	Project 5	Project 6
Urgency	Index 1	2.20	3.40	4.60	5.80	4.60	4.60
	Index 2	2.60	2.60	3.00	2.60	4.20	2.20
	Index 3	3.00	2.20	4.20	3.00	4.20	2.20
	Index 4	4.60	4.60	5.00	5.80	5.40	5.40
	Index 5	5.00	4.20	5.00	5.80	6.20	5.80
	Index 6	3.00	3.80	5.00	5.80	5.80	4.60
	Index 7	5.00	5.40	4.20	4.60	3.40	5.80
	Index 8	3.40	5.40	5.80	5.80	5.00	5.00
	Index 9	5.00	2.60	4.20	6.20	6.20	5.80
Possibility	Index 1	6.20	6.20	4.20	4.60	5.40	6.60
	Index 2	7.00	7.00	6.60	6.60	7.00	7.00
	Index 3	5.80	5.40	3.40	3.40	3.80	5.80
	Index 4	4.60	4.60	4.20	6.20	4.60	5.80
	Index 5	5.40	4.60	5.00	6.20	5.00	6.20
	Index 6	5.40	5.40	4.20	4.60	5.40	5.40
	Index 7	4.60	4.20	3.00	2.60	3.00	4.20

$$\max_{1 \leq k \leq s} \{\sigma_i^k\} = \sigma_i^{k^*}$$

We can determine the grey classes to which the three indices of project 1 belong: judged.

$$\max_{1 \leq k \leq 3} \{\sigma_1^k\} = \sigma_1^2 = 0.601$$

$$\max_{1 \leq k \leq 3} \{\rho_1^\kappa\} = \rho_1^2 = 0.536$$

$$\max_{1 \leq k \leq 3} \{\theta_1^\kappa\} = \theta_1^2 = 0.536$$

Table 6.14 Comprehensive Evaluation of Key Technology in Social Development Field

		Project 7	*Project 8*	*Project 9*	*Project 10*
Urgency	Index 1	3.00	1.00	5.80	5.00
	Index 2	5.33	4.67	5.00	5.00
	Index 3	5.33	4.00	4.60	5.33
	Index 4	5.00	4.67	4.60	5.67
	Index 5	3.67	3.33	5.00	4.33
	Index 6	3.33	2.67	4.60	4.00
	Index 7	3.00	3.33	3.80	2.67
	Index 8	4.33	3.33	6.60	6.00
	Index 9	4.00	5.33	7.00	7.00
Possibility	Index 1	4.33	5.33	5.80	4.00
	Index 2	7.00	6.67	6.20	6.67
	Index 3	4.33	4.00	4.60	4.33
	Index 4	4.33	5.00	5.40	5.67
	Index 5	5.00	5.00	5.80	5.67
	Index 6	5.33	5.67	4.20	4.67
	Index 7	3.00	3.33	2.60	5.33

Thus, the urgency, possibility, and comprehensive indices of project 1 fall into the medium class. The evaluations of other key technologies were conducted in the same way, and the 50 projects were ordered by their comprehensiveness indices; results are shown in Table 6.21.

Table 6.21 indicates that the superior, medium, and inferior classes represent 26, 24, and 50%, respectively, of the 50 projects. The six electronic information projects and four social development projects recommended represent 33 and 50% of superior and medium classes, respectively, and thus show superb technology cooperation potential. The modern agriculture, modern equipment manufacturing, new materials, new energy sources and energy saving projects show a certain hierarchy. Among nine new materials projects, four were categorized as superior

Table 6.15 Comprehensive Evaluation of Key Technology in Biotechnology and Innovative Therapeutics

		Project 11	Project 12	Project 13	Project 14	Project 15	Project 16	Project 17
Urgency	Index 1	7.00	7.00	5.80	2.20	5.80	5.80	5.80
	Index 2	4.20	5.00	3.80	1.80	3.00	4.20	1.80
	Index 3	4.20	3.00	2.20	1.80	3.00	5.00	1.80
	Index 4	5.00	6.20	5.80	3.00	3.00	4.20	3.80
	Index 5	4.60	5.00	4.60	3.80	4.60	4.20	4.60
	Index 6	5.40	3.80	4.20	1.40	4.20	4.60	2.60
	Index 7	2.20	1.80	3.80	5.40	3.80	3.40	3.40
	Index 8	5.40	5.00	3.80	3.80	3.00	4.60	3.80
	Index 9	3.80	3.80	3.00	3.40	2.60	2.60	2.60
Possibility	Index 1	5.40	5.00	6.20	6.20	5.00	5.80	5.40
	Index 2	6.60	7.00	5.80	7.00	6.60	7.00	5.40
	Index 3	5.80	6.20	3.80	5.40	5.00	5.40	5.40
	Index 4	5.80	6.20	4.20	3.00	3.80	4.60	3.80
	Index 5	5.00	5.40	3.40	4.20	5.00	5.80	4.20
	Index 6	4.60	4.60	3.80	5.80	6.20	4.60	5.00
	Index 7	5.00	5.00	3.40	6.60	5.40	4.60	5.00

Table 6.16 Comprehensive Evaluation of Key Technology in Modern Agriculture

		Project 18	Project 19	Project 20	Project 21	Project 22	Project 23	Project 24	Project 25
Urgency	Index 1	1.00	4.60	1.00	3.40	2.20	1.00	1.00	5.80
	Index 2	2.20	4.20	3.80	4.20	3.80	2.60	4.20	4.20
	Index 3	2.20	3.00	3.00	3.80	4.20	2.20	3.40	3.80
	Index 4	5.40	5.00	5.40	5.40	4.20	4.60	4.60	6.60
	Index 5	3.40	3.80	4.60	4.60	4.20	3.40	4.60	5.00
	Index 6	2.20	3.00	4.60	4.20	3.80	2.60	4.20	5.00
	Index 7	4.60	4.60	3.00	3.80	4.20	5.80	4.60	3.00
	Index 8	3.80	3.80	4.60	5.80	6.20	4.20	6.20	5.80
	Index 9	3.80	4.20	3.80	4.60	4.20	4.60	4.60	3.40
Possibility	Index 1	7.00	6.20	7.00	6.60	5.80	6.60	6.60	6.60
	Index 2	7.00	7.00	6.60	6.20	6.20	6.60	6.20	6.60
	Index 3	7.00	6.20	6.60	6.20	4.60	3.40	6.20	5.40
	Index 4	3.40	4.60	5.40	5.40	4.60	3.40	4.60	5.00
	Index 5	5.00	5.40	5.40	5.80	5.40	3.40	4.60	5.00
	Index 6	5.40	5.40	6.20	6.60	5.40	5.00	5.40	6.20
	Index 7	6.60	5.80	5.40	5.80	5.80	4.20	5.80	5.00

Table 6.17 Comprehensive Evaluation of Key Technology in Modern Equipment Manufacturing

		Project 26	Project 27	Project 28	Project 29	Project 30	Project 31	Project 32	Project 33	Project 34
Urgency	Index 1	7.00	2.50	1.00	2.50	2.50	1.00	1.00	5.80	1.00
	Index 2	5.00	3.00	3.00	3.00	4.50	3.50	3.00	3.00	2.50
	Index 3	4.50	3.00	2.50	3.50	3.50	3.50	3.00	3.00	2.50
	Index 4	5.00	6.00	6.00	4.50	5.00	5.00	5.00	6.60	6.00
	Index 5	4.50	4.50	3.50	4.00	4.50	5.00	4.50	4.60	5.00
	Index 6	3.00	3.00	2.00	2.00	2.50	4.00	3.00	4.60	3.00
	Index 7	4.50	5.50	6.00	4.00	5.50	5.00	4.00	2.20	5.00
	Index 8	6.50	6.00	3.50	4.00	4.00	4.50	5.00	5.80	4.50
	Index 9	4.50	4.00	2.00	6.50	1.50	4.00	5.00	3.80	4.00
Possibility	Index 1	4.50	6.50	7.00	6.50	5.50	6.50	6.50	6.20	6.50
	Index 2	7.00	7.00	7.00	4.00	7.00	7.00	7.00	7.00	6.50
	Index 3	5.50	6.00	5.50	4.00	5.50	5.50	6.00	7.00	6.00
	Index 4	5.00	6.00	2.00	4.00	3.50	5.50	4.50	5.80	3.50
	Index 5	5.50	5.50	4.50	3.50	3.50	5.50	3.50	5.00	4.00
	Index 6	6.00	6.00	6.50	5.00	5.00	6.50	6.50	5.40	5.50
	Index 7	5.00	5.00	6.50	3.50	5.00	5.00	5.00	5.40	4.50

Table 6.18 Comprehensive Evaluation of Key Technology in New Materials

		Project 35	Project 36	Project 37	Project 38	Project 39	Project 40	Project 41	Project 42	Project 43
Urgency	Index 1	1.00	1.00	1.00	7.00	1.00	1.00	7.00	1.00	7.00
	Index 2	5.00	5.00	3.00	5.00	1.00	2.00	1.67	3.00	1.67
	Index 3	3.00	3.00	3.00	3.00	1.00	2.00	1.67	3.00	1.00
	Index 4	5.00	5.00	3.00	3.00	7.00	6.00	5.00	3.00	5.67
	Index 5	5.00	3.00	3.00	5.00	3.00	5.00	7.00	5.00	5.67
	Index 6	1.00	3.00	3.00	3.00	1.00	3.00	5.67	3.00	5.67
	Index 7	5.00	5.00	5.00	5.00	7.00	6.00	3.67	5.00	4.33
	Index 8	5.00	7.00	5.00	7.00	5.00	5.00	5.00	7.00	6.33
	Index 9	5.00	5.00	3.00	3.00	5.00	4.00	5.67	1.00	7.00
Possibility	Index 1	7.00	7.00	7.00	7.00	7.00	6.00	7.00	7.00	6.33
	Index 2	7.00	7.00	7.00	7.00	7.00	7.00	7.00	7.00	6.33
	Index 3	3.00	3.00	3.00	3.00	7.00	5.00	1.67	5.00	3.67
	Index 4	3.00	5.00	3.00	3.00	7.00	6.00	7.00	3.00	5.67
	Index 5	3.00	5.00	3.00	3.00	7.00	6.00	7.00	3.00	5.67
	Index 6	6.00	5.00	7.00	7.00	7.00	5.00	5.67	7.00	4.33
	Index 7	5.00	3.00	5.00	1.00	7.00	4.00	3.67	3.00	5.00

Table 6.19 Comprehensive Evaluation of Key Technology in New Energy and Energy Saving

		Project 44	Project 45	Project 46	Project 47	Project 48	Project 49	Project 50
Urgency	Index 1	4.60	4.60	7.00	1.00	5.80	5.80	2.50
	Index 2	3.00	3.00	3.80	3.00	3.00	1.80	1.50
	Index 3	2.60	2.60	3.40	3.00	2.60	2.20	1.50
	Index 4	4.60	4.60	5.40	3.00	5.40	5.40	6.50
	Index 5	5.00	5.00	5.00	3.00	5.00	5.80	5.00
	Index 6	4.60	4.60	4.60	3.00	5.00	5.40	3.50
	Index 7	5.00	5.00	4.20	5.00	4.20	5.40	5.00
	Index 8	5.00	5.00	5.80	5.00	4.60	5.40	4.00
	Index 9	7.00	7.00	7.00	3.00	7.00	6.20	5.00
Possibility	Index 1	6.20	6.20	5.40	7.00	6.20	5.80	6.00
	Index 2	7.00	7.00	7.00	7.00	7.00	6.60	7.00
	Index 3	4.60	4.60	4.20	3.00	4.20	5.00	5.00
	Index 4	5.00	5.00	5.40	3.00	5.40	5.00	4.50
	Index 5	4.20	4.20	4.60	3.00	5.00	6.20	5.00
	Index 6	6.20	5.80	5.80	7.00	5.40	5.80	6.50
	Index 7	3.80	4.20	4.20	5.00	3.80	5.40	5.50

Table 6.20 Clustering Coefficients of Regional Cooperative Key Technology

Project	Urgency Index			Possibility Index			Comprehensive Index		
	Inferior	Medium	Superior	Inferior	Medium	Superior	Inferior	Medium	Superior
1	0.364	0.601	0.035	0.096	0.536	0.368	0.216	0.565	0.219
2	0.416	0.444	0.140	0.192	0.534	0.274	0.292	0.494	0.214
3	0.048	0.638	0.278	0.600	0.340	0.060	0.353	0.474	0.158
4	0.040	0.332	0.628	0.332	0.248	0.420	0.201	0.286	0.513
5	0.021	0.362	0.541	0.258	0.546	0.196	0.152	0.464	0.351
6	0.152	0.466	0.382	0.048	0.304	0.648	0.095	0.377	0.529
7	0.297	0.492	0.212	0.370	0.473	0.157	0.337	0.482	0.181
8	0.470	0.320	0.148	0.155	0.642	0.203	0.296	0.497	0.179
9	0.063	0.329	0.590	0.278	0.430	0.292	0.182	0.385	0.426
10	0.121	0.271	0.608	0.202	0.426	0.372	0.166	0.356	0.478
11	0.171	0.321	0.432	0.068	0.656	0.276	0.114	0.506	0.346
12	0.153	0.392	0.455	0.068	0.514	0.418	0.106	0.459	0.435
13	0.261	0.401	0.298	0.676	0.240	0.084	0.490	0.312	0.180
14	0.787	0.164	0.049	0.324	0.238	0.438	0.532	0.205	0.264
15	0.433	0.421	0.146	0.180	0.566	0.254	0.294	0.501	0.206

16	0.325	0.278	0.357	0.164	0.514	0.322	0.236	0.408	0.338
17	0.610	0.255	0.135	0.384	0.566	0.050	0.485	0.427	0.088
18	0.589	0.334	0.077	0.180	0.306	0.514	0.363	0.319	0.318
19	0.219	0.615	0.126	0.072	0.490	0.438	0.138	0.546	0.298
20	0.293	0.499	0.168	0.000	0.374	0.626	0.131	0.430	0.421
21	0.112	0.532	0.280	0.000	0.368	0.632	0.050	0.442	0.474
22	0.378	0.315	0.231	0.094	0.642	0.264	0.221	0.495	0.249
23	0.509	0.415	0.076	0.496	0.332	0.172	0.502	0.369	0.129
24	0.266	0.436	0.240	0.144	0.542	0.314	0.199	0.495	0.281
25	0.131	0.250	0.543	0.000	0.636	0.364	0.059	0.463	0.444
26	0.130	0.383	0.465	0.070	0.543	0.388	0.097	0.471	0.422
27	0.273	0.405	0.323	0.000	0.300	0.700	0.122	0.347	0.531
28	0.580	0.210	0.210	0.090	0.203	0.528	0.310	0.206	0.385
29	0.380	0.468	0.134	0.560	0.260	0.180	0.479	0.353	0.159
30	0.360	0.443	0.150	0.360	0.478	0.163	0.360	0.462	0.157
31	0.210	0.628	0.115	0.000	0.418	0.583	0.094	0.512	0.373
32	0.250	0.685	0.065	0.270	0.300	0.430	0.261	0.473	0.266

(Continued)

Table 6.20 (Continued) Clustering Coefficients of Regional Cooperative Key Technology

Project	Urgency Index			Possibility Index			Comprehensive Index		
	Inferior	Medium	Superior	Inferior	Medium	Superior	Inferior	Medium	Superior
34	0.360	0.433	0.208	0.390	0.315	0.295	0.377	0.368	0.256
35	0.260	0.540	0.200	0.470	0.120	0.410	0.376	0.308	0.316
36	0.295	0.440	0.265	0.230	0.530	0.240	0.259	0.490	0.251
37	0.555	0.345	0.100	0.470	0.120	0.410	0.508	0.221	0.271
38	0.315	0.270	0.415	0.590	0.000	0.410	0.467	0.121	0.412
39	0.540	0.185	0.275	0.000	0.000	1.000	0.242	0.083	0.675
40	0.455	0.270	0.275	0.060	0.410	0.530	0.237	0.347	0.416
41	0.202	0.362	0.437	0.190	0.097	0.713	0.195	0.216	0.589
42	0.465	0.370	0.165	0.480	0.110	0.410	0.473	0.227	0.300
43	0.190	0.153	0.657	0.187	0.447	0.367	0.188	0.315	0.497
44	0.092	0.610	0.298	0.238	0.408	0.354	0.173	0.499	0.329
45	0.092	0.610	0.298	0.214	0.466	0.320	0.159	0.531	0.310
46	0.000	0.401	0.541	0.164	0.500	0.336	0.091	0.456	0.428
47	0.555	0.345	0.100	0.470	0.120	0.410	0.508	0.221	0.271
48	0.036	0.552	0.412	0.116	0.560	0.324	0.080	0.556	0.364
49	0.172	0.276	0.552	0.000	0.544	0.456	0.077	0.424	0.499
50	0.330	0.495	0.175	0.090	0.540	0.370	0.198	0.520	0.283

Table 6.21   Order of Regional International Cooperative Key Technologies

Project Name	Class	Order	Project Name	Class	Order
New material 5	Superior	1	Electronic information 2	Medium	26
New material 7	Superior	2	Modern agriculture 2	Medium	27
Modern equipment manufacturing 2	Superior	3	Social development 1	Medium	28
Electronic information 6	Superior	4	Electronic information 3	Medium	29
Electronic information 4	Superior	5	Modern equipment manufacturing 7	Medium	30
New energy sources and energy saving 6	Superior	6	Modern equipment manufacturing 1	Medium	31
New materials9	Superior	7	Electronic information 5	Medium	32
Modern equipment manufacturing 8	Superior	8	Modern agriculture 8	Medium	33
Social development 4	Superior	9	Modern equipment manufacturing 5	Medium	34
Modern agriculture 4	Superior	10	Biotechnology and innovative medicines 2	Medium	35
Social development 3	Superior	11	New energy sources and energy saving 3	Medium	36
New materials 6	Superior	12	Modern agriculture 3	Medium	37
Modern equipment Manufacturing 3	Superior	13	Biotechnology and innovative medicines 6	Medium	38

(Continued)

Table 6.21 (Continued)   Order of Regional International Cooperative Key Technologies

Project Name	Class	Order	Project Name	Class	Order
Electronic information 1	Medium	14	Biotechnology and innovative medicines 4	Inferior	39
New energy sources and energy saving 5	Medium	15	New materials 3	Inferior	40
Modern agriculture 2	Medium	16	New energy sources and energy saving 4	Inferior	41
New energy sources and energy saving 2	Medium	17	Modern agriculture 6	Inferior	42
New energy sources and energy saving 7	Medium	18	Biotechnology and innovative medicines 3	Inferior	43
Modern equipment manufacturing 6	Medium	19	Biotechnology and innovative medicines 7	Inferior	44
Biotechnology and innovative medicines 1	Medium	20	Modern equipment manufacturing 4	Inferior	45
Biotechnology and innovative medicines 5	Medium	21	New materials 8	Inferior	46
New energy sources and energy saving 1	Medium	22	New materials 4	Inferior	47
Social development 2	Medium	23	Modern equipment manufacturing 9	Inferior	48
Modern agriculture 5	Medium	24	New materials 1	Inferior	49
Modern agriculture 7	Medium	25	Modern agriculture 1	Inferior	50

and four as inferior. Thus, resource allocation could be improved by choosing key technologies for international cooperation from this group. The nine biotechnology and innovative medicines projects revealed somewhat less potential than other project fields; 43% of the projects fell into the inferior class of regional cooperative key technology projects.

References

[1] Gu Jifa. A Summary of Evaluation Methods. Scientific Decision and Systems Engineering. Beijing: China Press of Science and Technology, 1990.

[2] Wang Yingluo. Systems Engineering. Beijing: China Machine Press, 2008.

[3] Thomas L. Saaty and Luis G. Vargas. Models, Methods, Concepts and Applications of the Analytic Hierarchy Process. Norwell, MA: Kluwer Academic, 2001.

[4] Edgeworth, Francis. The statistics of examinations. *Journal of the Royal Statistics Society,* 51, 599–684, 1888.

[5] Spearman, Charles. Correlations of sums and differences. *British Journal of Psychology*, 5, 417–426, 1913.

[6] A. Charnes, W.W. Cooper, and E. Rhodes. Measuring the efficiency of decision making units. *European Journal of Operational Research,* 2, 429–444, 1978.

[7] Thomas R. Sexton, Richard H. Silkman, and Andrew J. Hogan. Data Envelopment Analysis: Critique and Extensions. San Francisco: Jossey-Bass, 1986.

[8] Quanling Wei. Data Envelopment Analysis. Beijing: Science Press, 2004.

[9] Xiaojian Chen and Liang Liang. Systems Evaluation Methods and Application. Hefei: Press of China University of Science and Technology, 1993.

[10] Shoukang Qin Principles and Applications of Synthetical Evaluation. Beijing: Publishing House of Electronics Industry, 2003.

[11] Hong Gao, Huanchen Wang, and Yuqing Tian. Comprehensive evaluation of achievements and benefits of science and technology in Yellow River sector. *Systems Engineering Theory and Practice,* 19, 114–119, 1999.

[12] Yang Yu and Yijun Li. Research on giving weight for performance indicators based on the multi-strategy method. *Systems Engineering Theory and Practice,* 23, 9–15, 2003.

[13] Guotai Chi, Yanqiu Cheng, and Lijun Wang. The society evaluation model based on grey clustering and its empirical study of cities. *Chinese Journal of Management Science*, 18, 185–192, 2010.

[14] Wei Meng Daqun Zhang, and Wenbin Liu. A study of multi-level DEA models and applications. *Chinese Journal of Management Science*, 16, 148–154, 2008.

[15] Jianqiang Wang and Shichang Ren. Grey random multi-criteria decision-making approach based on expected value. *Control and Decision,* 24, 39–43, 2009.

[16] Yantai Chen, Guohong Chen, and Meijuan Li. Classification and research advancement of comprehensive evaluation methods. *Journal of Management Sciences in China,* 7, 70–79, 2004.

[17] Sifeng Liu, Dang Yaoguo, Fang Zhigeng, and Xie Naiming. Grey Systems Theory and Its Applications, 5th ed. Beijing: Science Press, 2010.

[18] Sifeng Liu and Yi Lin. Grey Systems: Theory and Applications. Heidelberg: Springer, 2011.

[19] Sifeng Liu and Yi Lin. Grey Information: Theory and Practical Applications. London: Springer London Ltd., 2006.

[20] Lirong Jian, Sifeng Liu, and Yi Lin. Hybrid Rough Sets and Applications in Uncertain Decision-Making. Boca Raton, FL: Taylor & Francis Group, 2011.

[21] Yaoguo Dang, Sifeng Liu, and Yuhong Wang. Optimization of Regional Industrial Structures and Applications. Boca Raton, FL: Taylor & Francis Group, 2011.

[22] Sifeng Liu and Li Jie. Analysis of distribution structure and efficiency of funds employed for science and technology in China. *Science and Technology and Economy,* 18, 26–28, 2005.

[23] Hecheng Wu and Sifeng Liu. Evaluation of R&D relative efficiency of different areas in China based on improved DEA model. *R&D Management,* 19, 108–112, 2007.

[24] Jiefang Wang, Sifeng Liu, and Muyuan Liu. Grey incidence analysis models with incomplete information based on cross evaluation. *Systems Engineering Theory and Practice,* 30, 732–737, 2010.

[25] Ke Zhangke and Sifeng Liu. Extended clusters of grey incidences for panel data and their application. *Systems Engineering Theory and Practice,* 30, 1253–1259, 2010.

[26] Sifeng Liu, Naiming Xie, and Jeffrey Forrest. On new models of grey incidence analysis based on visual angle of similarity and nearness. *Systems Engineering Theory and Practice,* 30, 881–887, 2010.

[27] Sifeng Liu, Wenfeng Yuan, and Keqin Sheng. Multi-attribute intelligent grey target decision model. *Control and Decision,* 25, 1159–1163, 2010.

[28] Sifeng Liu. On index system and mathematical model for evaluation of scientific and technical strength. *Kybernetes,* 35, 1256–1264, 2006.

[29] Sifeng Liu, Li Bingjun, et al. Study of mathematical model and indices for the synthetic evaluation of regional leading industry. *Chinese Journal of Management Science,* 6, 8–13, 1998.

[30] Sifeng Liu, Xuewen Tang, Chaoqing Yuan, and Yaoguo Dang. Study on degree of orderliness of industrial structure. *Economic Perspectives,* 5, 53–56, 2004.

[31] Sifeng Liu and Ling Yang. On Evaluation, Precaution and Adjustment. Zhengzhou: Henan People's Press, 1994.

[32] Sifeng Liu and Yongda Zhu. Study on triangular model and indexes in synthetic evaluation of regional economy. *Transactions of the Chinese Society of Agricultural Engineering,* 9, 8–13, 1993.

[33] Sifeng Liu and Naiming Xie. A new grey evaluation method based on reformative triangular whitenization weight function. *Journal of Systems Engineering,* 2011.

[34] Chaoqing Yuan, Sifeng Liu, and Junlong Wu. Research on energy-saving effect of technological progress based on Cobb-Douglas production function. *Energy Policy,* 37, 2842–2846, 2009.

[35] Chaoqing Yuan, Sifeng Liu, and Naiming Xie. The impact on Chinese economic growth and energy consumption of the global financial crisis: an input–output analysis. *Energy,* 35, 1805–1812, 2010.

[36] Chaoqing Yuan, Sifeng Liu, and Junlong Wu. The relationship among energy prices and energy consumption in China. *Energy Policy,* 38, 197–207, 2010.

[37] Chaoqing Yuan, Sifeng Liu, and Junlong Wu. Research on energy-saving effect of technological progress based on Cobb-Douglas production function. *Energy Policy,* 38, 2611–2617, 2010.

[38] Chaoqing Yuan, Sifeng Liu, and Zhigeng Fang. The relation between Chinese economic development and energy consumption in the different periods. *Energy Policy,* 38, 5189–5198, 2010.

[39] Chaoqing Yuan, Sifeng Liu, Zhigeng Fang. Research on the energy-saving effect of energy policies in China: 1982–2006. *Energy Policy,* 37, 2475–2480, 2009.

[40] National Bureau of Statistics of China. China Statistical Yearbook (1996). China Statistics Press, 1996.

[41] National Bureau of Statistics of China. China Statistical Yearbook (1997). China Statistics Press, 1997.

[42] National Bureau of Statistics of China. China Statistical Yearbook (1998). China Statistics Press, 1998.

[43] National Bureau of Statistics of China. China Statistical Yearbook (1999). China Statistics Press, 1999.

[44] National Bureau of Statistics of China. China Statistical Yearbook (2000). China Statistical Press, 2000.

[45] National Bureau of Statistics of China. China Statistical Yearbook (2001). China Statistics Press, 2001.

[46] National Bureau of Statistics of China. China Statistical Yearbook (2002). China Statistics Press, 2002.

[47] National Bureau of Statistics of China. China Statistical Yearbook (2003). China Statistical Press, 2003.

[48] National Bureau of Statistics of China. China Statistical Yearbook (2004). China Statistics Press, 2004.

[49] National Bureau of Statistics of China. China Statistical Yearbook (2005). China Statistics Press, 2005.

[50] National Bureau of Statistics of China. China Statistical Yearbook (2006). China Statistics Press, 2006.

[51] National Bureau of Statistics of China. China Statistical Yearbook (2007). China Statistics Press, 2007.

[52] National Bureau of Statistics of China. China Statistical Yearbook (2008). China Statistics Press, 2008.

[53] National Bureau of Statistics of China. China Statistical Yearbook (2009). China Statistics Press, 2009.

[54] National Bureau of Statistics of China. China Statistical Yearbook (2010). China Statistics Press, 2010.

[55] Jiangsu Provincial Government. Eleventh Five-Year Technology Development Plan Report in Jiangsu Province, Jiangsu Statistics Press, 2006.

[56] David Romer. Advanced Macroeconomics. Shanghai University of Finance & Economics Press, 2001, pp. 5–17.

[57] Wang Xiaolu and Fang Gang. Sustainability of China's Economic Growth-Retrospect and Prospect at the Turn of Century. Bejing: Economic Science Press, 2000, pp. 57–65.

[58] Sifeng Liu and Yi Lin. Grey Systems: Theory and Applications. Heidelberg: Springer, 2011, pp. 151–172.

Index

A

absolute degree of grey incidence 46
accumulative total contribution rate 26
affirmative answer without provisory conditions
        13
AHP 32
Alex Faickney Osborne 5
analytic hierarchy process 32
analyst 7
anonymity 9
Aristotle 3
average random consistency index 36
Average Weakened Buffer Operator
        (AWBO) 147

B

before and after comparison 25
benefit type objective 61
brain storming method 5

C

characteristic equation 21
center-point triangular whitenization
        function 57
Charles Spearman 2
Check Sissies 42
close degree of grey incidence 53
clustering coefficient 208
comparison method 25
complementary AHP model 32
comprehensive clustering ccoefficient 57
consistency index 36
consistency ratio 36
conditional answer 13
confidence probability index 10

confidence factor 10
contribution rate of a principal component 21
controlled brainstorming 6
correlation matrix 21
cost type objective 61
countermeasure 58
countermeasure set 58
critical value 62
C²R design 40
C²R model 39

D

data envelopment analysis 39
DEA 39
DEA effective 42
decision making units 39
deductive answer 13
Delphi method 8
derived Delphi method 9
desirable countermeasure 59
desirable situation 59
direct brainstorming 6
discrete parameters 22
distribution parameter 22
DMU 39
doubtful brainstorming 6
dual problem 42

E

effect mapping 59
effect measures 61
effect value of situation 59
efficiency evaluation index 40
encouraging observation 6
end-point triangular whitenization function 56
eigenvalue 21

eigenvector 21
eigenvector method 24
Elasticity Coefficient 95
energy intensity 141
energy policy 142
energy saving 141
evaluation 1
evaluation period 84
event 59
event list 9
event set 59
expert of deduction 7

F

feasibility study 73
Financial Benefit Cost Ratio (FBCR), 86
Financial Internal Rate of Return
      (FIRR) 86
Financial Investment Payback
      Period (Pt) 86
Financial Net Present Value (FNPV) 86
Francis Edgeworth 1

G

general steps of NGT 5
general steps of qualitative evaluation 5
generalized grey incidence model 46
GM(1,1) 82
grey clustering with fixed weight 208
grey cluster evaluation model 54
grey linear programming model 154
grey target 59
grey target of two-dimensional
      decision making 60

H

Hierarchical structure model 32
hit the target 60
hit the bull's-eye 62,65
horizontal Logic relation 29

I

idea producer 7
indicator quantified 21
industrial restructuring 151
industrial structure 153
initial image 50
initial operator 50

input surplus 44
input vector 41
international cooperation project 189

J

judgment matrix 35

K

key technology 191

L

Lao Tzu 4
least squares method 23
length of a sequence 47
logical framework approach 26
lower effect measure 61
Lower measure whitenization weight
      function 205
lower threshold value 59

M

mapping variable 22
Markov Chain Model 174
methodology scholar 7
moderate measure whitenization weight
      function 205
moderate value type objective 61

N

nearness degree of grey incidence 53
negative deviation 22
non-Archimedean infinitesimal 42
non-DEA effective 41
normalization 22
Nominal Group Technique 5
nonfossil energy 173
non-sequence-based system for multilevel index
      19
k-th subclass 57
normative condition 62

O

observation encouragement approach 6
one-dimensional grey target of decision making
      59
optimum countermeasure 64

optimum event 64
optimum situation 42–64
optimum value 42
output deficit 44
output vector 41

P

patent 105
principal component 21
positive deviation 22
positive reciprocal matrix 34
possibility index 200
postevaluation 30
preevaluation 68
principal component analysis 20
priority order 208
process evaluation 61
production function 142

Q

quadratic programming model 174
qualitative evaluation 3
questionnaires 13
questionnaire surveying goals and means 13

R

relative degree of grey incidence 50
Research and Development(R&D) 105

S

satisfactory effect 59
s-dimensional grey target of decision making 59
scientific and technological activities 105
scientific and technological inputs 105
scientific and technological outputs 105
scientific and technological strength 121
sensitivity analysis 90
sequence-based system for multilevel index 19
similitude degree of grey incidence 47

single-layer system 19
situation 59
situation set 59
slack variable 118
standard deviation coefficient 22
steps of system evaluation 2
strategy observation method 6
symmetry 22
synthetic degree of grey incidence 52
synthetic effect measure 58
systems analysis 9

T

Taoist 4
total Investment 73
traffic flow 77
triangular whitenization function 54
typical whitenization weight function 204

U

urgency index 209
uniform effect measure 60
upper effect measure 61
upper measure whitenization weight function of 206
upper threshold effect value 62

V

vertical logic 28

W

weakly DEA effective 43
whitenization weight function 46
with and without comparison 25

Z

zero-starting point operator 46
zigzagged line 46